FLORA OF THE GUIANAS

Edited by

M.J. JANSEN-JACOBS

Series A: Phanerogams

Fascicle 24

7. HERNANDIACEAE
(A.S.J. van Proosdij)

8. CHLORANTHACEAE
(C.A. Todzia)

9. PIPERACEAE
(A.R.A. Görts-van Rijn)

including
Wood and Timber
(A.M.W. Mennega)
(J. Koek-Noorman)

2007
Royal Botanic Gardens, Kew

Contents

7. HERNANDIACEAE

by

André S.J. van Proosdij[1]

Trees, (climbing) shrubs or lianas. Leaves not stipulate, petiolate, often peltate, alternate; blade simple or up to 3-5-lobed or divided into 3-5 leaflets; veination prominent, pinnate, palmate or camptodromous; oil cells, sometimes cystoliths and glandular hairs present. Inflorescences axillary, pseudo-terminal, rarely terminal panicles, usually much branched, terminal-flower generally absent, flowers in cincinni or dichasia; bracts and bracteoles present or absent. Flowers bisexual, unisexual, or by abortion unisexual (then plants monoecious, very rarely dioecious), regular, epigynous; perianth segments 3-10, in 1 or 2 whorls, quincuncial, (sub-)imbricate or valvate; stamens 1-7, in 1 whorl, alternating with (inner) whorl of perianth segments, usually free, in *Hernandia* with 2 somewhat connate, swollen glands on back, anthers basifixed, 2-locular, introrse, dehiscing by valves opening to the top or to the side, connective linear, pollen grains inaperturate, spinulose; interstaminal glands very small or absent; ovary present or rudimentary, unilocular, ovule 1, anatropous, pendulous, style 1, simple, with a groove decurrent along entire length. Fruits dry, indehiscent drupes, ovoid or ellipsoid, rarely globose, longitudinally ribbed, in *Hernandia* included in cupule; seed 1, endosperm absent, cotyledons 2, large, with apical radicle.

Distribution: Pantropical, most frequent in coastal regions, 4 genera and 62 species; in the Neotropics 3 genera with 24 species: *Gyrocarpus* (2 species), *Hernandia* (7 species) and *Sparattanthelium* (15 species); in the Guianas 4 species in 2 genera; *Hernandia* (1 species) and *Sparattanthelium* (3 species).

LITERATURE

Espejo, A. 1992. Hernandiaceae. Flora de Veracruz 67: 1-22.

Grenand, P., *et al.* 1987. Pharmacopées traditionnelles en Guyane, Hernandiaceae. p. 245-247.

Kostermans, A.J.G.H. 1937. Hernandiaceae. In A.A. Pulle, Flora of Suriname 2(1): 338-344.

[1] Nationaal Herbarium Nederland, Utrecht University branch, Heidelberglaan 2, 3584 CS Utrecht, The Netherlands.

Kubitzki, K. 1969. Monographie der Hernandiaceen. Bot. Jahrb. Syst. 89: 78-209.
Kubitzki, K. 1976. Hernandiaceae. In J. Lanjouw & A.L. Stoffers, Flora of Suriname, Additions and Corrections 2(2): 485-486.
Kubitzki, K. 1982. Hernandiaceae. In Z.L. de Febres & J.A. Steyermark, Flora de Venezuela 4(2): 324.
Lemée, A.M.V. 1955-56. Hernandiacées, Flore de la Guyane Française, 1: 658-659 and 4: 36.
Meisner, C.D.F. 1866. Lauraceae (Sparattanthelium) and Hernandiaceae. In C.F.P. von Martius, Flora Brasiliensis 5(2): 291-294 and 297-300.
Miller, J.S. & P.E. Berry. 1999. Hernandiaceae. In J.A. Steyermark *et al.*, Flora of the Venezuelan Guayana 5: 592-593.
Mori, S.A. & J.L. Brown. 2002. Hernandiaceae. In S.A. Mori *et al.*, Guide to the Vascular Plants of Central French Guiana. Mem. New York Bot. Gard. 76(2): 344-347.
Pax, F. 1889. Hernandiaceae. In H.G.A. Engler & K.A.E. Prantl, Die natürlichen Pflanzenfamilien 3(2): 126-129.

KEY TO THE GENERA

1 Trees; inflorescences with cincinni, bracts present; flowers unisexual, stigma broadly peltate; fruits included in a cupule . . . . *1. Hernandia*
Lianas or shrubs; inflorescences with dichasia, bracts in general absent; flowers bisexual, stigma knob-like; fruits not included in a cupule . . . . *2. Sparattanthelium*

1. **HERNANDIA** L., Sp. Pl. 2: 981. 1753.
Type: H. sonora L.

Trees, rarely shrubs, in general evergreen (rarely deciduous). Leaves simple (rarely 3-5-lobed), often peltate, cystoliths absent. Inflorescences consisting of a branched axis and flowers in cincinni; peduncles elongate (rarely short); cincinni with 3 flowers (rarely 2 or 1): 2 (1) lateral male flower(s) and 1 central, subsessile female flower (rarely bisexual) on branches of the first (rarely second) order, or on main axis; upper parts of branches somewhat tawny or white pubescent, rarely glabrous; bracts generally present; bracteoles of male flowers 4, ± equal, free, partly including flower buds from cincinnus, bracteoles of female flower 2, connate, forming fleshy, campanulate cupule. Flower buds ovoid, rarely globose. Flowers bisexual or unisexual, then monoecious, rarely dioecious; perianth segments free, minutely pubescent on both sides, outer segments quincuncial or imbricate, inner segments narrower and imbricate or valvate. Male flowers: 6-12 perianth segments; stamen filaments short, each with 2 yellow, somewhat stalked, glandular appendages, free or united; pollen 90-160 μm, spinose; staminodes absent; pistil rudimentary. Female flowers: 8-12 perianth segments; stamens absent, glands 4-5(-10-12), free or connate; ovary laterally somewhat compressed, style sigmoid or straight, moderately thick or thinner towards base, stigma mostly diagonally placed, large, fleshy, margin fimbriate. Drupes black, ovoid or ellipsoid, ribbed/striped, included in cupule; cotyledons free and rather thick, or connate and ruminate. Dispersal of fruits zoöchorous or hydrochorous.

Distribution: Pantropical, 26 species; most frequent in coastal regions in Central America, the Guianas, West Indies, West Africa, Indomalaysia, Pacific Islands.

1. **Hernandia guianensis** Aubl., Hist. Pl. Guiane 2: 848, t. 329. 1775. Type: French Guiana, Kaw Distr., property of M. Boutin, Aublet s.n. (holotype BM, isotypes NY, G). – Fig. 1 A-H

H. sonora sensu Kosterm. (1937), non L. (1753).
H. sonora sensu Lemée (1955), non L. (1753).

Tree up to 30 m high, with buttresses; bark glabrous, shining greyish-white; wood white, with a soft aroma of *Apium*. Branches fragile and breaking easily. Leaves simple; petiole terete, 6-14(-22) cm long, fleshy at base; blade papery to somewhat leathery, longitudinally folded, elliptic-ovate, 11-20(-30) x 6-14(-25) cm, margin entire, apex 1-2(-3) cm acuminate, base rounded, subcordate or acute, not peltate, only leaves of young plants up to 10 mm peltate, green; at base sub-5-veined, 2-4 pairs of higher secondary veins, veins below loosely pubescent with short glandular and longer straight hairs, otherwise leaves glabrous on both sides. Inflorescence axillary or pseudo-terminal, pubescent, conspicuously ash-grey; peduncle (6-)10-16 cm long; cincinni on secondary or tertiary branches; bracts 0-2, underneath ramifications, variable in shape; bracteoles of male flowers oblong, 5-8 x 3-4 mm. Flower buds ellipsoid or obovoid, 3-5 x 2-3.5 mm. Flowers unisexual. Male flowers: pedicels 2-3 mm long; perianth segments 6, in 2 whorls, whitish, elliptic, 5-6(-7) x 2-2.5 mm, apex and base rounded, densely pubescent; stamens 3, little shorter than perianth, orange-yellow, filaments free, loosely pubescent, anther valves opening to sides; each stamen with 2 glabrous glands. Female flower: pedicel 0-1 mm long; cupule broadly cup-shaped, not constricted, ca. 2 x 2 mm, entire; perianth as in male flower; style little shorter than perianth, thin at base, broader towards stigma, recurved at top, lanate. Fruit pedicel up to 5 mm long. Fruit

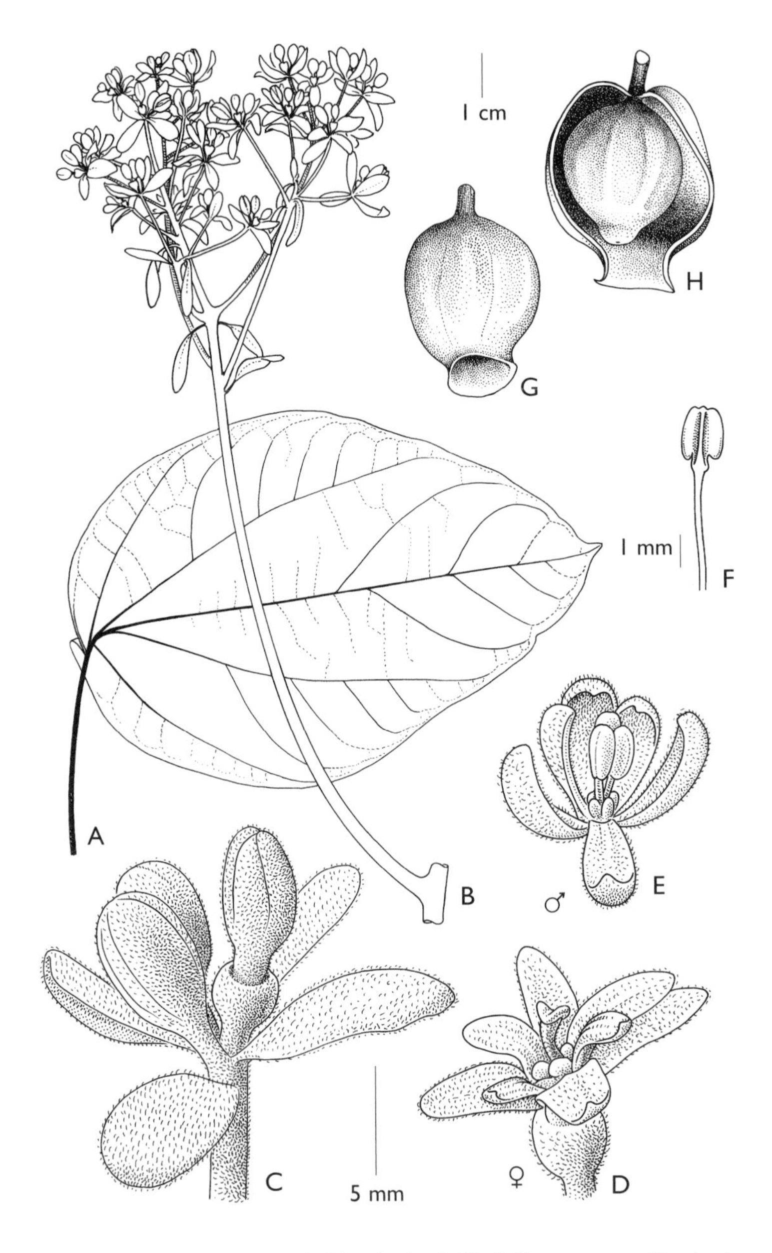

Fig. 1. *Hernandia guianensis* Aubl.: A, leaf; B, inflorescence; C, cincinnus consisting of 1 central female flower and 2 lateral male flowers; D, female flower with young cupule; E, male flower; F, stamen with valves opened to the sides; G, fruit cupule; H, opened cupule, showing drupe (A-B, Polak *et al*. 72; C-F, van Andel *et al*. 2646; G-H, Prévost 3500). Drawing by H. Rypkema.

cupule succulent, red spotted, up to 5.5 x 4 cm, with an almost circular, slightly constricted hole. Drupe brown, ovoid, ca. 3 x 2.4 cm, with 8 longitudinal slightly prominent ribs/grooves, glabrous, apex almost flat, with rounded umbo; cotyledons ruminate, connate.

Distribution: In coastal regions from Trinidad and Venezuela to the Guianas and the Amazon region; occasional to frequent in rain forest and secondary forest, as well as on shell-ridges, in swamp forest or along flooded river banks, also found on laterite soil and in mixed forest on light brown sandy loam, lowland; 26 collections studied (GU: 8; SU: 3; FG: 13).

Selected specimens: Guyana: Berbice R., S of New Dageraad, Maas *et al.* 5562 (NY, U); Port Kaituma, Polak *et al.* 72 (B, NY, U). Suriname: Coppename R., between Coppename and Coronie, Lanjouw & Lindeman 1547 (K, NY, U); Saramaca R., klein Moho, Hamburg, BBS 112 (U). French Guiana: Cayenne, garden of centre ORSTOM, Prévost 3500 (CAY, P, U, US); Oyapock R., near St. Georges, Cremers 7020 (U).

Uses: The light and strong wood is traded and used as timber. The Galibis and the Bush-negros use the wood the way we use a tinder/fire lighter. The flowers and the fruit, especially the kernel are used by the Garipons (coming from Pará, Brazil) and by some people from Cayenne to make purgative emulsions. The Palikur use emulsions from the flowers and the fruit against whooping cough, for which the local name is 'maoksikan', which means 'cry of the howler monkey'.

Vernacular names: Guyana: takariwa (Akawaio); hoahoa/howa howa, kaioballi, takariwa (Arawak); jack-in-the-box (Creole); kurokuroru (Warrau). Suriname: popolika. French Guiana: bois banane (based on the white and poreuze wood like the false stem of a banana), mirobolan (the fruit) (Creole); ajowa, haiwowa (Carib); maoksi adudu (maoksi = howler monkey, adudu = goitrous, to the form of the mature cupule) (Palikur); waliwowo (describing the noise of the wind in the enlarged cupule) (Wayampi).

Phenology: Flowering in Jan. to Jul. and Sep. to Dec.; fruiting in Mar. to May.

Note: Many specimens of *H. guianensis* from the Guianas have been wrongly identified as *H. sonora* L. which is known to occur in the Antilles, the most southern place being Trinidad. The latter differs from *H. guianensis* in having larger leaves which are always 1.2-2.0 cm peltate. The drupe of *H. sonora* has a contracted narrow base and a clearly constricted umbo, whereas *H. guianensis* has a broad base and a non-constricted umbo. The drupe of *H. sonora* has much more prominent grooves.

2. **SPARATTANTHELIUM** Mart., Flora 24 (2, Beibl.): 40. 1841.
Type: S. tupiniquinorum Mart.

Small evergreen trees, shrubs, or lianas climbing with lateral, often recurved short-shoots. Branches terete, glabrous or pubescent or with stellate-congested hairs on young branches, finally glabrate. Leaves simple, more dense towards end of branchlets; blades margin entire, apex acuminate, deep green, laxly pubescent or glabrous above, dense and whitish lanuginose or more or less glabrous below; in general 3-veined or 3-pliveined, 2-4 pairs of secondary veins, rarely more or less pinnately veined, intersecondary veins parallel reticulate, all veins below prominulous; cystoliths present. Inflorescences numerous, axillary, just under top of branch or terminal, spreading, dichasial, 8-13 times branched, terminal flower of final dichasia usually lacking; pedunculate; bracts and bracteoles in general absent, sometimes basal ramifications with leaf-like bracts; branches moderately to strongly nodose, sometimes dense and whitish or laxly pubescent. Flower buds ovoid or ellipsoid, greyish white pubescent. Flowers bisexual, or functional unisexuall; perianth lobes 4-8, in one whorl, valvate when 4-5, irregular sub-imbricate when 6-8, more or less persistent, outside greyish white pubescent, inside green to dark green, glabrous; stamens 4-5(-7), alternating with perianth lobes, filaments short, glabrous, anthers elongate, glabrous, valves opening from base to top, finally horizontally spreading, pale yellow, pollen 19-30 μm, in white heaps; staminodes and glandular appendages absent; style cylindrical, thick, with appressed hairs, stigma knob-like, small. Infructescence broad, many branched panicle, hanging, bearing few fruits, nodose, silver-white, rarely dark or brown; fruit pedicels discoid at top. Drupes silver whitish, rarely dark or brownish, often compressed, glabrous, striped, sarcocarp thin, endocarp stony, coriaceous or woody; cotyledons narrowly folded.

Distribution: 15 species in the Neotropics, from Mexico to Brazil; 3 species in the Guianas.

KEY TO THE SPECIES

1 Leaves pinnately veined; short-shoots hook-formed, up to 2 cm long, leafless; infructescence and drupes dark brown or black . . . . *1. S. aruakorum*
Leaves 3-pliveined; short-shoots sometimes present, hook-formed, 4-16 cm long, usually with leaves; infructescence and drupes silvery white or silvery grey 2

2 Leaf surface glabrous or (laxly) pubescent below; fruit pedicel 0.9-1.7 cm long . . . . *2. S. guianense*
Leaf surface densely pubescent below; fruit pedicel 2.5-6 cm long . . . . *3. S. wonotoboense*

1. **Sparattanthelium aruakorum** Tutin, J. Bot. 78: 249. 1940. Type: Guyana, Bartica-Portaro road, Tutin 252 (holotype BM, isotypes K, U, US).

Liana. Stem dark green. Short-shoots up to 2 cm long, recurved, hook-formed, compressed at base, striate, without leaves, glabrous, when young laxly pubescent. Leaves: petiole 1-3.5 cm long, glabrous; blade slightly coriaceous, narrowly oblong to ovate, 6-12(-13) x 2.5-4.5 cm, margin flat, apex 0.5-1 cm acuminate, base rounded or obtuse, when dry olive-green to dark green above, slightly shining, glabrous, below somewhat paler olive or grey, glabrous; at base 3-veined, appearing to be pinnately veined, secondary veins 4-6 pairs, in a right angle with main vein, all veins above and below glabrous or laxly pubescent, on both sides prominulous. Inflorescence axillary or pseudo-terminal, 8-10 times branched, up to 13.5 cm long; peduncle 4.5-6(-7) cm long, dark, glabrous, branchlets loosely minutely pubescent; pedicels thin, greyish white pubescent. Flowers: perianth lobes 5, rarely 4, narrowly obovate/ovate/oblong to ligulate, 1.5-1.8 x 0.8-1 mm, more or less hooded; stamens 4, filaments 0.5 mm long, anthers ca. 1 mm long; style 0.6-0.8 mm long. Infructescence up to 22 cm long, nodose, dark brown to black; fruit pedicel 4.5-5.5 cm long. Drupe dark brown or black, ellipsoid, 1.7-1.8 x 0.6-0.7 cm, compressed, inconspicuously 6-ribbed.

Distribution: Guyana; on white sand in clearings and wallaba (*Eperua*) bush, lowland; only 2 fertile collections seen, both from the Bartica-Potaro road: the type and Forest Dept. 5586 = Fanshawe 2787 (U) and 1 sterile collection from Suriname: Brokopondo Distr., W of Village Brokopondo, van Donselaar 3754 (U) (GU: 2; SU: 1).

Phenology: Flowering in Jun., fruiting in Nov.

2. **Sparattanthelium guianense** Sandwith, Bull. Misc. Inform. Kew 1932: 225. 1932. Type: Guyana, Essequibo R., Moraballi Cr., near Bartica, Sandwith 470 (holotype K, isotypes NY, U).

Liana. Short-shoots 5-13 cm long, recurved, leaf-bearing and often hooked at top; ultimate branchlets striate, densely minutely pubescent, lower ones brown, glabrous. Leaves: petiole (0.5-)1.5-2.5(-3.2) cm long, densely pubescent, glabrescent; blade thick papery to slightly coriaceous, ovate-, obovate- or oblong-elliptic, 4-16.5 x 1.7-8 cm, margin slightly revolute, apex 0.3-1.5 cm pointed, base rounded or obtuse, when dry dark olive-green above, subglabrous, below somewhat paler, glabrous or laxly pubescent; at base 3-veined, secondary veins 1-2 pairs, intersecondary veins more or less parallel, at right angle with main vein, veins pubescent above when young, later glabrous, below densely, later laxly minutely pubescent. Inflorescence 9-11 times branched, up to 10-12 cm long; peduncle 5-9 cm long, ash grey, minutely pubescent; pedicel thin, ash grey, minutely pubescent. Flowers: perianth lobes 4-5, oblong, 1.5-2 x 0.5-0.8(-1) mm; stamens 4-5, filaments 0.3 mm long, anthers ca. 1 mm long; style 1.8-2 mm long. Infructescence up to 20 cm long, slightly nodose, silvery grey when mature; fruit pedicel 0.9-1.7 cm long. Drupe silvery grey, elongate-ellipsoid, 1.5-1.7 x 0.6-0.7 cm, longitudinally slightly compressed, inconspicuously 5-7-ribbed, rugose/ somewhat wrinkled.

Distribution: Guyana; mixed forest on brown or white sand, lowland (GU: 4).

Selected specimens: Guyana: Demerara R., Jenman 4889 (K); Sandhills, Forest Dept. 3938 = Fanshawe 1202 (K, NY); Pomeroon-Supenaam, Kurushi Cr., Hoffman *et al.* 2764 (U, US).

Phenology: Flowering and fruiting in Feb., May, Oct.

3. **Sparattanthelium wonotoboense** Kosterm., Meded. Bot. Mus. Herb. Rijks Univ. Utrecht 25: 44. 1936, as '*wonotoboensis*'. Type: Suriname, Wonotobo, Corantijne R., BW 3120 (holotype U, isotype NY).
– Fig. 2 A-C

S. botocudorum sensu Pulle, Recueil Trav. Bot. Néerl. 9: 138. 1912, non Mart. (1841).
S. botocudorum sensu Kosterm. (1937), non Mart. (1841).
S. botocudorum Mart. var. *uncigerum* Meisn. in Mart. Fl. Bras. 5(2): 293. 1866. – *S. uncigerum* (Meisn.) Kubitzki, Bot. Jahrb. Syst. 89-1: 196. 1969. Type: French Guiana, Karouany, Sagot 1218 (holotype BR, isotypes P, U), syn. nov.
S. macusiorum A.C. Sm., Lloydia 2: 181. 1939. Type: Guyana, Kanuku Mts., NW slopes, Moku Cr. (Takutu tributary), A.C. Smith 3390 (holotype NY, isotypes B, K, NY, U).
S. melinonii Baill. ex Lemée, Fl. Guyan. Franç. 4: 36. 1956 (nom. inval.).

Liana. Branches terete, smooth, striate; short-shoots thick, terete, 4-16 (-20) cm long, recurved, hooked or arcuate at top, strongly compressed at base, glabrous or densely pubescent when young, darkening, striate.

Fig. 2. *Sparattanthelium wonotoboense* Kosterm.: A, flowering branch; B, infructescence; C, flower at anthesis (A, Cremers 9581; B, Smith 3390; C, Jansen-Jacobs *et al*. 3180). Drawing by H. Rypkema.

Leaves: petiole thin, 0.8-2.5(-4) cm long, densely minutely pubescent to more or less glabrous; blade papery to slightly coriaceous, ovate-, obovate- or oblong-elliptic, 4-11(-14) x 2-5(-7) cm, margin somewhat revolute, apex 0.3-1 cm acuminate, base rounded or obtuse, sometimes shortly acute, above brownish green to (dark) olive, glabrous or laxly pubescent, sometimes a bit shining, below paler or greyish green to ash-grey, densely puberulent, trichomes 0.1-0.7 mm long; sub-3-pliveined, secondary veins 1-4 pairs, all veins above glabrous to pubescent, below glabrous to densely pubescent. Inflorescences axillary, 8-12(-13) times branched, up to 10-14 cm long; peduncle 3-8 cm long, grey pubescent or glabrous, branchlets terete, greyish white pubescent; pedicels slender, grey pubescent. Flowers: perianth lobes 4, elliptic or oblong, 1.5-2 x 0.5-1 mm, smudgy red; stamens 4, filaments 0.5 mm long, anthers 1 mm long; style thick, 1.2-1.8 mm long. Infructescence 12-24 cm long, nodose to strongly nodose, silver-white when mature, pubescent or glabrous; branchlets slender; fruit pedicel slender, 2.5-6(-6.5) cm long. Drupe silver-white, ellipsoid, 1.3-1.6 x 0.6-0.7 cm, longitudinally compressed, acute, inconspicuously 6(-8)-ribbed, somewhat wrinkled.

Distribution: Venezuela, the Guianas, Brazil; forest near granite rocks, savanna forest, non-flooded moist forest, high forest along rivers and wallaba (*Eperua*) forest, 0-400 m elev.; 30 collections studied (GU: 7; SU: 7; FG: 12).

Selected specimens: Guyana: Rupununi Distr., Kuyuwini Landing, Kuyuwini R., Jansen-Jacobs *et al.* 3180 (B, NY, U, US). Suriname: Nickerie Distr., Kabalebo Dam project, between km 29 & 30, Lindeman & Görts-van Rijn *et al.* 57 (NY, U, US). French Guiana: Saül, airport road, Mori *et al.* 23044 (B, CAY, U, US); Ile de Cayenne, River S., Oldeman B-3890 (CAY, NY, P, U).

Vernacular names: Suriname: oneka (Carib). French Guiana: wilalãkãyewi (Wayampi).

Phenology: Flowering in Feb., Apr.-Jul., Sep and Oct.; fruiting in Jan., Apr.-Aug., Oct. and Nov.

Note: In 1969 Kubitzki raised *S. botocudorum* var. *uncigerum* to the rank of species. The differences he mentioned between *S. uncigerum* and *S. wonotoboense* are that the former has glabrous young branches and petioles, slightly coriaceous leaves, 0.3-0.4 mm long hairs on the lower surface of the leaf, buds 0.6-1.0 mm in diameter, strongly nodose infructescence and drupes 1.5-1.6 x 0.6-0.7 cm. The latter has densely villose young branches and petioles, papery leaves, 0.5-0.8 mm long hairs on the lower surface of the leaf, buds 1.2 mm in diameter, slender and nodose infructescence and drupes 1.2-1.5 x 0.4 cm.

Studying the collections I noticed the great similarity between the two species. Most characters such as leaf texture, shape and size, indument of branches, petioles and inflorescences, characters of the flowers and fruits showed to vary in such a way that the ranges in which these characters occur in one species, do overlap those of the other species. Therefore, no character or combination of characters can be found that clearly distinguishes the 2 species.

Typical specimens of *S. wonotoboense* were collected in Suriname and Guyana, but not in French Guiana. Typical *S. uncigerum* specimens were found in French Guiana and Suriname, but not in Guyana. Intermediary forms were collected in all of the three countries.

Conclusion: *S. uncigerum* and *S. wonotoboense* as well as all the intermediate forms belong to one single variable species: *S. wonotoboense.*

8. CHLORANTHACEAE

by

CAROL A. TODZIA[2]

Trees, shrubs or herbs. Leaves opposite, decussate, simple; petioles connate at base, stipulates present, petiolar; blades pinnately veined, usually glabrous, margins dentate. Inflorescences capitate, spicate or racemose, axillary or terminal. Flowers small, unisexual (monoecious or dioecious) or bisexual, without perianth or with 3-lobed calyx, subtended by 1-3 bracts or ebracteate; stamens 1-3, in bisexual flowers adnate to ovary, anthers 2- or 4-locular, linear to oblong, opening with longitudinal slits, connective often expanded or extended. Pistillate and bisexual flowers epigynous, hemi-epigynous or naked; carpel 1, stigma sessile, ovule 1, orthotropous. Fruit a berry with a hard seed coat or a drupe with a fragile, stony endocarp (*Hedyosmum*); seeds with well-developed starchy endosperm, embryo small with 2 minute cotyledons.

Distribution: About 75 species in 4 genera, all found in the moist tropics of the world; in the Neotropics 1 genus with 44 species; in the Guianas 1 species.

LITERATURE

Todzia, C.A. 1988. Chloranthaceae: Hedyosmum. Flora Neotropica 48: 1-139.

Todzia, C.A. 1993. Chloranthaceae. In K. Kubitzki, The Families and Genera of Vascular Plants 2: 281-287.

Todzia, C.A. 1998. Chloranthaceae. In J.A. Steyermark *et al.*, Flora of the Venezuelan Guayana 4: 198-201.

1. **Hedyosmum** Swartz, Prodr. 847. 1788.
   Type: H. nutans Swartz

Monoecious or dioecious aromatic shrubs or trees, rarely herbs, often with prop roots; wood white, usually soft; stems with persistent leaf sheaths or with encircling leaf sheath scars; nodes swollen. Leaves fleshy to coriaceous when fresh; margins dentate; petioles with expanded bases

[2] Plant Resources Center, The University of Texas, Austin, Texas 78712, U.S.A.

forming a connate sheath around stem; distal margin of leaf sheath with or without stipular appendages. Staminate inflorescences composed of a solitary spike or with several spikes on a racemose or paniculate axis, subtended by a pair of leafy bracts; spikes with 50-300 flowers. Staminate flowers consisting of a solitary, sessile, ebracteate stamen; stamens quadrangular to oblong, 4-locular, longitudinally dehiscent, anther connective extended with apex, flat, acute or acuminate. Pistillate inflorescences solitary, thryse-like, racemose, or paniculate, subtended by leafy bracts; flowers solitary or more often clustered into cymules. Pistillate flowers consisting of an ellipsoidal or trigonous ovary, with perianth adnate to ovary with 3 free or partially fused lobes at apex of ovary, subtended by a chartaceous or fleshy floral bract that often encloses flower. Fruit a drupe with a fleshy wall formed by fused perianth or multiple with connate floral bracts becoming fleshy and colored (white or purple); seeds small, brown or black, ellipsoidal or trigonous, smooth or minutely papillate.

Distribution: About 44 species, mostly at mid-elevations from Mexico to Paraguay; 1 species in SE Asia; in the Guianas 1 species.

1. **Hedyosmum tepuiense** Todzia, Novon 3: 83. 1993. Type: Venezuela, Amazonas, Dept. Rio Negro, Cerro de la Neblina, Anderson 13389 (holotype TEX, isotypes MICH, NY, VEN). – Fig. 3

Dioecious aromatic tree or shrub, 2-16 m tall; young stems quadrate, older stems terete with tubular leaf bases persisting; internodes 3-6 cm long, glabrous. Free portion of petiole 0.7-2 cm long, glabrous to slightly floccose or strigose; petiolar sheaths 1.5-2.5 cm long, usually verrucose with two linear to fimbriate stipular appendages ca. 3 mm long; leaf blade elliptic, 7.5-19 x 2.3-10 cm, with long-acuminate tip 0.5-1.5 cm long, attenuate at base, margin serrate, drying scabrous; midvein impressed above, raised beneath, with floccose or strigose hairs. Staminate inflorescence 2.5-6.5 cm long, rachis with 2-3 nodes, each with 1-3 spikes; mature spike 0.6-1.5 cm long with 50-100 stamens; anthers 1-1.5 cm long terminated by an expanded apiculate connective. Pistillate inflorescence racemose or paniculate, 3-10 cm long with ca. 20 cymules; cymule with 2-5 pistillate flowers; pistillate flower slightly trigonous, 2-3 cm long, 1-2 mm broad; perianth lobes small, ca. 0.3 mm long; stigmas irregularly clavate, 1-2 mm long, 3-angled, papillose. Fruiting cymules white, irregularly globose, 4-9 mm diam.; seeds ellipsoidal, ca. 3 mm long, brown, smooth.

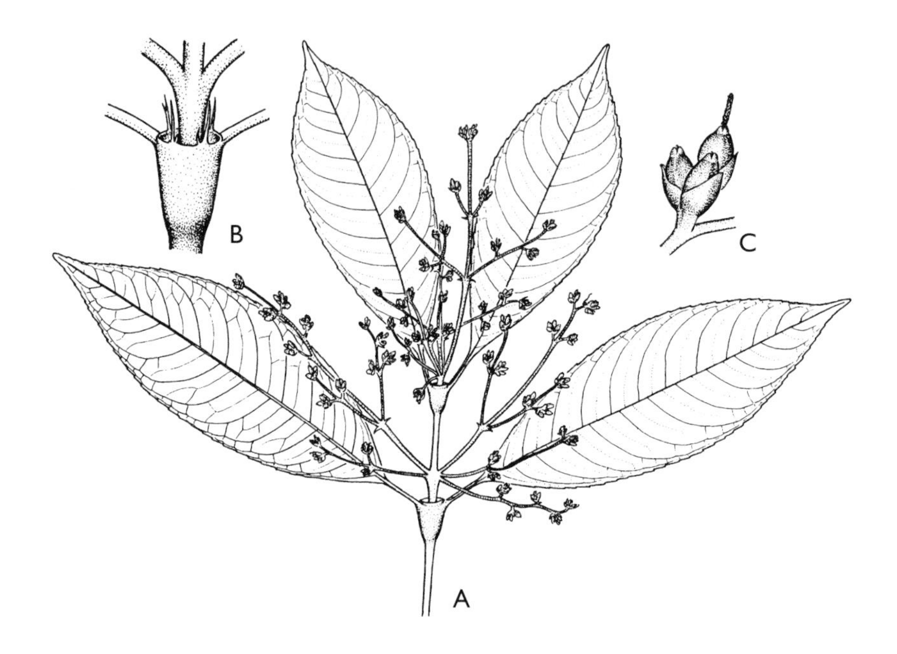

Fig. 3. *Hedyosmum tepuiense* Todzia: A, flowering branch; B, petiolar sheath with stipular appendages; C, pistillate cymule. Drawing by B. Manara; reproduced with permission of the Missouri Botanical Garden Press from J.A. Steyermark, P.E. Berry and B.K. Holst (eds.), Flora of the Venezuelan Guayana 4: 201, Fig. 142. 1998.

Distribution: Venezuelan Guayana and Pakaraima Mts. in Guyana; wet montane forest and riparian habitats, 1200-2600 m elev.; 20 collections studied (GU: 3).

Specimens examined: Guyana: Mt. Roraima, Hahn & Gopaul 5481a (TEX, US); Mt. Wokomung, Boom & Samuels 9063, 9200 (NY, TEX).

9. PIPERACEAE

by

ARA GÖRTS-VAN RIJN[3]

Herbs, shrubs or treelets, sometimes scandent, rarely lianas. Leaves simple, alternate, opposite or in whorls; stipules absent or present; prophyll often present, sometimes enclosing shoot-apex. Inflorescences terminal, axillary, or leaf-opposed spikes with minute flowers, solitary or few together, or occasionally spikes arranged in panicles or umbels. Flowers sessile, usually bisexual, without calyx or corolla, subtended by a floral bract; stamens 1-6; ovary sessile or stipitate, 1-locular with 1 ovule, stigmas 1-4. Fruits small drupes or nuts.
Fruits are said to be dispersed by bats or epizoochorous.

Distribution: A pantropical family with more than 3000 species, the number of genera is ca. 8; in the Guianas 2 genera with 88 species.

Note: The descriptions are mainly based on herbarium specimens. *In vivo* specimens may differ in leaf texture and the measures may be different. When too few collections were available, data from Steyermark, Callejas & Steyermark, or Mathieu were used.
In the species descriptions some characters are given only, when they are important to distinguish morphological similar taxa, like for instance: anther dehiscence, indument of the spike rachis, the prophyll and stipules. To give them in all descriptions would make the treatment too lengthy. In the paragraphs specimens examined only those duplicates have been noted that I have actually seen.

[3] Nationaal Herbarium Nederland, Utrecht University branch, Heidelberglaan 2, 3584 CS Utrecht, The Netherlands.

Acknowledgements. My sincere thanks go to those who enabled me to produce this part of the Flora. Firstly thanks to the curators of the herbaria in C, CAY, G, K, L, NY, U, US, who provided the many collections. Hendrik Rypkema, artist at Utrecht, made most of the drawings, unless stated otherwise. Guido Mathieu (GENT) gave me information on living *Peperomia,* his website (www.peperomia.net/repertory.asp) was an extremely useful tool for finding data and illustrations. In the final phase I could not have done without the help of Maarten C.M. Christenhusz, then student at Utrecht. Ricardo Callejas Posada (HUA) and Eric Tepe (MU) sent many useful comments on the draft version. I am very grateful to my Utrecht colleagues. Marion Jansen-Jacobs and Jan Lindeman checked the draft versions extremely thoroughly on inconsistencies and errors in keys and text. Gea Zijlstra studied the nomenclature and corrected references and names of several species that appeared to be incorrect or illegal. Renske Ek kindly worked on finalizing the list of collections studied.

LITERATURE

Bornstein, A.J. 1991. The Piperaceae in the Southeastern United States. J. Arnold Arbor., Suppl. Ser. 1: 349-366.

Burger, W.C. 1971. Flora Costaricensis, Piperaceae. Fieldiana, Bot. 35: 5-227.

Candolle, A.C.P. de. 1869. Piperaceae. In A.L.P.P. de Candolle, Prodromus systematis naturalis regni vegetabilis 16(1): 235/65-471.

Görts-van Rijn, A.R.A. 2002. Piperaceae. In S.A. Mori *et al.*, Guide to the Vascular Plants of Central French Guiana. Mem. New York Bot. Gard. 76(2): 574-584.

Görts-van Rijn, A.R.A. & R. Callejas Posada. 2005. Three new species of Piper (Piperaceae) from the Guianas. Blumea 50: 367-373.

Howard, R.A. 1973. Notes on the Piperaceae of the Lesser Antilles. J. Arnold Arbor. 54: 377-411.

Jaramillo, M.A. & P. Manos. 2001. Phylogeny and patterns of floral diversity in the genus Piper (Piperaceae). Amer. J. Bot. 88: 706-716.

Kramer, K.U. & A.R.A. Görts-van Rijn. 1968. Piperaceae. In A.A. Pulle & J. Lanjouw, Flora of Suriname, Additions and Corrections 1(2): 415-421.

Lemée, A.M.V. 1955-1956. Flore de la Guyane Française. Pipéracées. 1: 470-492. 1955; 4: 25-27. 1956.

Mathieu, G. 2001-2006. The Internet Peperomia Reference, www.peperomia.net/.

Miquel, F.A.W. 1843-1844. Systema Piperacearum. p. 1-304. 1843; p. 305-575. 1844.

Stearn, W.T. & L.H.J. Williams, 1957. Martin's French Guiana plants and Rudge's «Plantarum Guianae Rariorum Icones». Bull. Jard. Bot. État 27: 243-265.

Steyermark, J.A. 1984. Piperaceae. Flora de Venezuela 2(2): 1-619.

Steyermark, J.A. & R. Callejas. 2002. Piperaceae. In J.A. Steyermark *et al.*, Flora of the Venezuelan Guayana 7: 681-738.

Tebbs, M.C. 1989. Revision of Piper (Piperaceae) in the New World. I. Review of characters and taxonomy of Piper section Macrostachys. Bull. Brit. Mus. (Nat. Hist.), Bot. 19: 117-158.

Tebbs, M.C. 1990. Revision of Piper (Piperaceae) in the New World. 2. The taxonomy of Piper section Churumayu. Bull. Brit. Mus. (Nat. Hist.), Bot. 20: 193-236.

Tebbs, M.C. 1993. Revision of Piper (Piperaceae) in the New World. 3. The taxonomy of Piper sections Lepianthes and Radula. Bull. Brit. Mus. (Nat. Hist.), Bot. 23: 1-50.

Trelease, W. & T.G. Yuncker. 1950. The Piperaceae of Northern South America. p. 1-838.

Verdcourt, B. 1996. Piperaceae. In R.M. Polhill, Flora of Tropical East Africa. 24 pp.
Yuncker, T.G. 1957. Piperaceae. In A.A. Pulle & J. Lanjouw, Flora of Suriname 1(2): 218-290.

KEY TO THE GENERA

1 Herbs, usually epiphytic or on rocks, occasionally terrestrial; nodes not thickened; leaves succulent, fleshy; stigma 1; floral bracts round-peltate, glabrous . . . . *1. Peperomia*
Herbs, subshrubs, shrubs or treelets, occasionally scandent, trailing, or lianas, terrestrial; nodes thickened; leaves not fleshy; stigmas (2-)3-4; floral bracts various, glabrous or pilose, often marginally fringed . . . . *2. Piper*

1. **PEPEROMIA** Ruiz & Pav., Fl. Peruv. Prodr. 8. 1794.
Type: P. secundiflora Ruiz & Pav.

Acrocarpidium Miq., Inst. Verslagen Meded. Kon. Ned. Inst. Wetensch. 1842: 198. 1843.
Type: A. nummulariifolium (Sw.) Miq. ('nummularifolium') (Piper nummulariifolium Sw.) ('nummularifolium')
Micropiper Miq., Comm. Phytogr. 32, 39. 1840.
Type: not designated

Small epiphytic, epilithic, or terrestrial, succulent herbs, creeping, prostrate, hanging, ascending or erect. Leaves alternate, opposite or in whorls or 3-7 per node; pinnately or palmately veined. Inflorescences terminal, leaf-opposed or axillary spikes, solitary or few together, occasionally paniculate; flowers numerous, loosely or less often densely arranged, sometimes partly sunken in the rachis; rachis glabrous (but hairy in *P. tetraphylla)*; floral bracts peltate, mainly rounded, glabrous, often glandular. Stamens 2, filaments short, anthers 2-locular; ovary with 1 style and stigma, or stigma sessile. Fruits minute nuts or drupes ("berries"), (sub)basely or laterally attached, sessile or somewhat stipitate when mature, occasionally somewhat sunken in rachis, smooth or sticky, globose, ellipsoid or cylindrical, apex obliquely scutellate or truncate, with or without a slender beak, stigma apical, central or at base of beak.

Distribution: Pantropical, ca. 1000 species, ca. 500 in the Neotropics; 30 species in the Guianas of which 4 endemics.

N o t e: According to Burger (1971: 7), the viscid layer of the fruit may in drying give rise to structures like the "pseudopedicel". It may also be effective in the formation of the beak that is present in several taxa. This beak, however, is also distinct in living collections (G. Mathieu pers. comm.).

KEY TO THE SPECIES

1 Leaves opposite or in whorls of 1-7 2
Leaves alternate 10

2 Leaf apex acute or somewhat acuminate 3
Leaf apex obtuse, rounded or emarginate 4

3 Stem glabrous; leaves glabrous; spikes 10-16 cm long; fruit with apical stigma . . . . *2. P. angustata*
Stem villous; leaves pubescent; spikes 2.5-15 cm long; fruit with subapical stigma . . . . *3. P. blanda*

4 Leaves opposite 5
Leaves in whorls of 1-7 6

5 Stem minutely hirtellous; leaf margin not ciliate; peduncle without bracts . . . . *4. P. delascioi*
Stem minutely pubescent; leaf margin ciliate; peduncle 2-bracteate . . . . *21. P. quadrangularis*

6 Stem glabrous . . . . *22. P. quadrifolia*
Stem variously pubescent 7

7 Fruit slightly stipitate at maturity; stigma apical . . . . . . *24. P. rhombea*
Fruit (sub)basely attached, not stipitate; stigma apical on short, thick style, or fruit with small beak and subapical stigma, or fruit apex oblique with subapical stigma 8

8 Rachis of spike pubescent; fruits sunken in rachis . . . *28. P. tetraphylla*
Rachis of spike glabrous; fruits (sub)basely attached, not sunken 9

9 Spikes to 15 cm long, terminal or axillary, often in groups of 2-5; leaf blade 0.2-0.6 mm wide, apex obtuse; fruit apex oblique with subapical stigma . . . . *6. P. galioides*
Spikes 2-7 cm long, terminal and solitary; leaf blade 0.6-2.3 cm wide; apex rounded to emarginate; fruit with short, thick style and apical stigma . . . . *14. P. maguirei*

10(1) Spikes in panicles . . . . *18. P. pernambucensis*
Spikes simple, solitary, 2-few together or bifurcate 11

11 Leaves distinctly peltate 12
Leaves basely attached or slightly peltate 14

12 Leaf apex rounded, blade with erect trichomes at upper surface; petiole attached centrally; fruit flat at apex . . . . *8. P. gracieana*
Leaf apex somewhat acute to acuminate, blade glabrous or minutely pubescent; petiole attached between base and leaf centre, 0.2-3 cm from base; fruit with slender beak 13

13 Stem and leaf blade minutely pubescent; petiole 3-11(-16) cm long . . . . *10. P. hernandiifolia*
Stem and leaf blade glabrous; petiole 1.5-6 cm long . . *29. P. transparens*

14(11) Leaf apex rounded or emarginate 15
Leaf apex acuminate, acute to obtuse 21

15 Stem and leaf blade glabrous; blade 2-14 cm long; spikes 2-18 cm long 16
Stem and leaf blade crisp-pubescent or minutely hirtellous; blade less than 2 cm long; spikes to 2 cm long 18

16 Petiole not grooved or winged; peduncle as long as or longer than spikes; spikes 2-5 cm long . . . . *9. P. haematolepis*
Petiole grooved or winged; peduncle as long as or shorter than spikes; spikes to 18 cm long 17

17 Leaf blade in vivo fleshy; floral bracts 5-7 mm in diam.; fruit subglobose . . . . *13. P. magnoliifolia*
Leaf blade in vivo coriaceous; floral bracts 2-4 mm in diam.; fruit ellipsoid . . . . *15. P. obtusifolia*

18 Leaf apex emarginate; stem glabrous, sparsely villous or minutely hirtellous 19
Leaf apex rounded; stem crisp-pubescent 20

19 Stem minutely hirtellous; leaf blade 0.5-0.8 x 0.5-1 cm; fruit crowned with persistent, conic style, stigma mammiform; above 2000 m elev. . . . . *4. P. delascioi*
Stem glabrous or sparsely villous; leaf blade 0.15-0.6 cm in diam.; fruit with apical stigma, not mammiform; from sealevel up to 600 m elev. . . . . *5. P. emarginella*

20 Leaf blade broadly elliptic to deltoid; peduncle nodose or with 1-2 (falling off early) bracts, often longer than flowering rachis; fruit ellipsoid . . . . *26. P. serpens*
Leaf blade almost circular, broadly elliptic or elliptic; peduncle not nodose, without bracts, usually shorter than flowering rachis; fruit globose to ovoid . . . . *25. P. rotundifolia*

21(14) Terrestrial herbs 22
Epiphytic or epilithic, occasionally terrestrial herbs 23

22 Leaf blade broadly elliptic to deltoid, 1-3.7 cm long; spikes terminal, to 6 cm long . . . . *17. P. pellucida*
Leaf blade ovate, to 12 cm long; spikes axillary, to 17 cm long . . . . *19. P. popayanensis*

23 Leaf base truncate, cord(ul)ate or rounded; spikes 1-5 cm long 24
Leaf base cuneate, acute to obtuse; spikes 1.5-30 cm long 25

24 Leaf blade 0.8-2.5 x 0.8-2.5 cm; spike axillary, 1-2.5 cm long, shorter than peduncle . . . . *23. P. reptans*
Leaf blade 2-3 x 2.7-7 cm; spike terminal, 1.5-5 cm long, about as long as peduncle . . . . *30. P. urocarpa*

25 Older leaf blades 4-20 cm long; spikes 4-30 cm long 26
Older leaf blades less than 4 cm long; spikes 1.5-7 cm long 30

26 Whole plant strongly glandular dotted (black when dried); petiole usually with 2 rows of hairs; peduncle glabrous . . . . *7. P. glabella*
Plants at most somewhat glandular, not black-dotted when dried; no lines of hairs on petiole; peduncle glabrous or minutely puberulous 27

27 Internodes of stem longitudinally ridged or winged; petiole deeply grooved, glabrous; fruits laterally attached, globose-ovoid . . . *1. P. alata*
Internodes not winged nor ridged; petiole narrowly winged or not, glabrous or sparsely, minutely puberulent; fruits basely attached, cylindrical or ellipsoid 28

28 Leaf blade narrowly elliptic to ovate, upper surface usually covered with waxy exudate forming star- or scale-like structures; spikes to 30 cm long; growing on ant's nests . . . . *12. P. macrostachya*
Leaf blade narrowly elliptic to obovate, upper surface not covered with waxy exudate; spikes to 18 cm long; not growing on ant's nest . . . 29

29(25) Leaf blade 1-3 cm wide, apex acuminate; fruits with short beak . . . . *11. P. lancifolia* subsp. *lancifolia*
Leaf blades 2.5-6 cm wide, apex at most acute; fruits tapering into a slender beak, which is slightly hooked at very tip . . . . *13. P. magnoliifolia*

30 Leaf blade 0.8-2.3 cm wide, often, or at least at veins, red below; spikes 5-7 cm long . . . . *20. P. purpurinervis*
Leaf blade less than 1 cm wide, green; spikes to 5.5 cm long 31

31 Leaf blade villous, apex obtuse; veins obvious; fruit sessile . . . . *16. P. ouabianae*
Leaf blade glabrous or shortly puberulous, apex obtuse and minutely emarginate with some hairs in the indentation; veins obsolete; fruit stipitate at maturity . . . . *27. P. tenella*

1. **Peperomia alata** Ruiz & Pav., Fl. Peruv. Chil. 1: 31, t. 48, f. b. 1798. – *Piper alatum* (Ruiz & Pav.) Vahl, Enum. Pl. 1: 342. 1804. Type locality: Peru, Pozuo; type not designated. – Fig. 4 A-B

Epiphytic, epilithic or terrestrial, glabrous herb. Stem ascending, 10-40 cm long; internodes winged or ridged below nodes, green or reddish. Leaves alternate; petiole 0.3-3 cm long, deeply grooved or winged; blade fleshy membranous or coriaceous, green above, often reddish brown beneath, elliptic or broadly elliptic or ovate, 4.5-13 x 2-5 cm, margin may be ciliate towards apex, apex acuminate, base acute, glabrous; 3-5(-7)-pli- or palmately veined. Inflorescence terminal or axillary, single or few together, erect; peduncle not slender, 0.3-1.5 cm long, green; spike 6-14 cm long, yellow; floral bracts rounded, glabrous, slightly glandular. Fruits laterally attached, globose-ovoid, reddish brown, apex oblique with subapical stigma.

Distribution: Mexico, C America, West Indies, Colombia, Venezuela, the Guianas, Brazil, Ecuador, and Peru; from sea level up to 1700 m elev.; 31 collections studied (GU: 16; SU: 15; FG 1).

Selected specimens: Guyana: Cuyuni-Mazaruni region, Koatse R., Pipoly 10799 (U, US); Hahn 5227 (U). Suriname: Nassau Mts., Marowijne R., Maguire 40704 (NY, U); Lely Mts., Lindeman & Stoffers *et al.* 127, 423 (NY, U). French Guiana: near Maripasoela, Schnell 11582 (P, U).

2. **Peperomia angustata** Kunth in Humb., Bonpl. & Kunth, Nov. Gen. Sp. ed. qu. 1: 68. 1816. – *Piper angustatum* (Kunth) Poir. in Lam., Encycl. Suppl. 4: 469. 1816. Type: Venezuela, Prov. Sucre, near Cumanacoa and San Fernando, Humboldt & Bonpland 1171 (holotype P, isotype B-W 745). – Fig. 5 A

Peperomia victoriana C. DC. in A. DC., Prodr. 16(1): 449. 1869. Type: Venezuela, Edo. Aragua, La Victoria, Fendler 1819 (holotype G-DC).

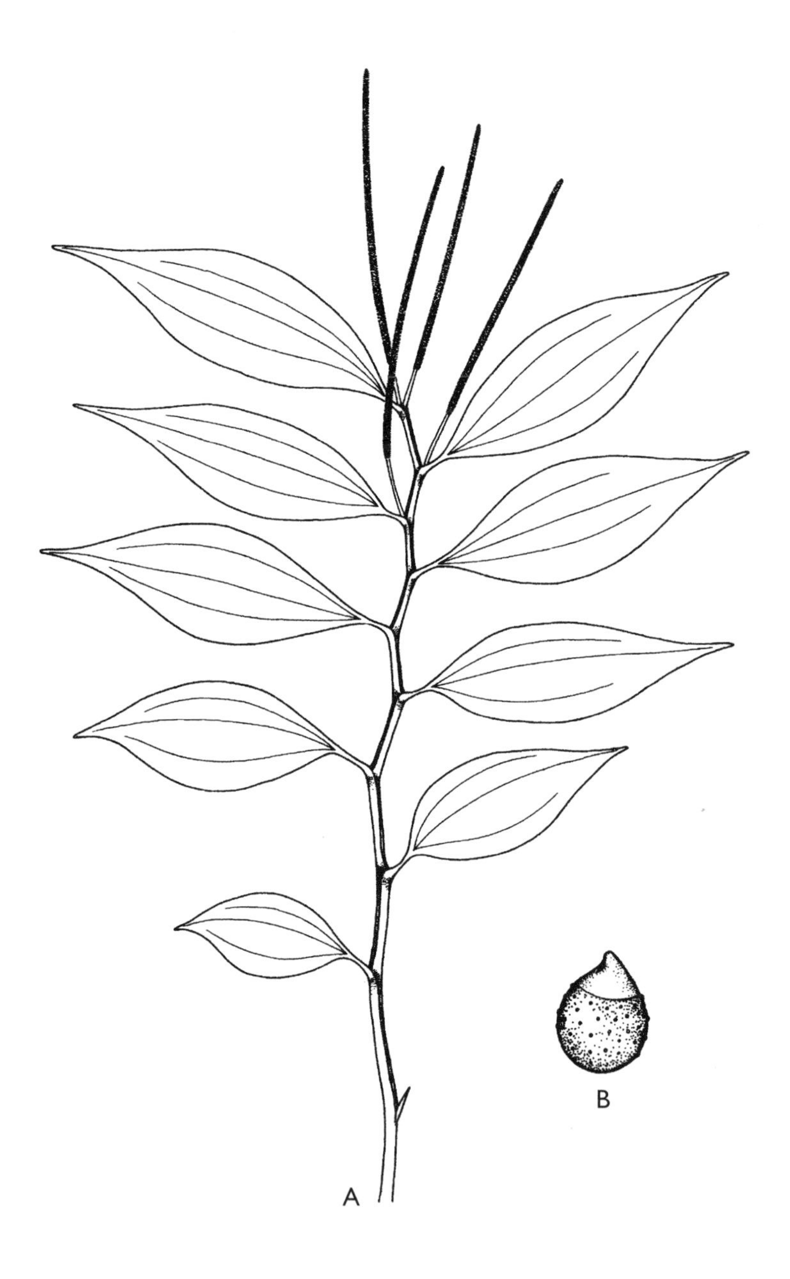

Fig. 4. *Peperomia alata* Ruiz & Pav.: A, habit; B, fruit. After drawing in Flora de Venezuela.

Terrestrial, epiphytic or epilithic herb, erect, creeping or ascending, rooting at nodes. Stem glabrous, 30-100 cm long, green or reddish. Leaves in whorls of (2-)3-4(-5), basely attached; petiole 0.2-1.5 cm long, glabrous; blade fleshy to coriaceous, rhombic to elliptic or oblanceolate, 2-6(-9) x 1-3(-4) cm, apex acute or acuminate, base acutish to cuneate, glabrous; palmately 3(-5)-veined. Inflorescence single, terminal, erect; peduncle slender, 2-3 cm long, glabrous or minutely pubescent, red; spike 10-16 cm long, moderately to densely flowered; floral bracts rounded to oblong, slightly glandular. Fruits basely attached, ovoid to globose, beaked, blackish, often verruculose, apex obliquely scutellate, stigma apical.

Distribution: C America, NE Colombia, Venezuela, Guyana, French Guiana, N Brazil and Peru; in forests, from sea level up to 2400 m elev.; 7 collections studied (GU: 4; FG: 4).

Selected specimens: Guyana: Cuyuwini R., A.C. Smith 2544 (K, U); Mt. Roraima, im Thurn 139 (BM, K). French Guiana: Approuague R., Kapiri Cr., Cremers 11570 (CAY, U); near Grand Croissant, 40 km N of Camopi, Feuillet 1173, 1213 (CAY, K, P, U); Aratai Cr. near junction with Approuague R., Mori *et al.* 25665 (CAY, U).

Note: The Guyanan collection Schomburgk s.n. from Mt. Parima has smaller leaves (1.5 x 0.7 cm) but in my opinion belongs to *Peperomia angustata*.

3. **Peperomia blanda** (Jacq.) Kunth in Humb., Bonpl. & Kunth, Nov. Gen. Sp. ed. qu. 1: 67. 1816. – *Piper blandum* Jacq., Collectanea 3: 211. 1791. Type: Venezuela, Caracas, Jacquin s.n. (isotype B-W 747). – Fig. 5 B-C

Terrestrial, epiphytic or epilithic herb, erect or ascending from rooting base. Stem villous, often fuscous, green or reddish, 10-20(-60) cm long. Leaves opposite or in whorls of 3-7 (sometimes dimorphic), basely attached; petiole 0.2-1(-1.8) cm long, pubescent; blade slightly fleshy, elliptic, narrowly elliptic, or obovate, often rhombic, 2-8 x 0.8-4 cm, margin ciliate, apex acute or somewhat acuminate, base acute, pubescent usually on both sides; palmately 3(-5)-veined. Inflorescence terminal or axillary, single or few together; peduncle 0.7-3.2 cm long, pubescent or crisp-pubescent; spike 2.5-15 cm long, green, densely to laxly flowered; floral bracts rounded, glandular. Fruits subbasely attached, globose to ovoid, more or less verruculose, brown, apex oblique with subapical stigma.

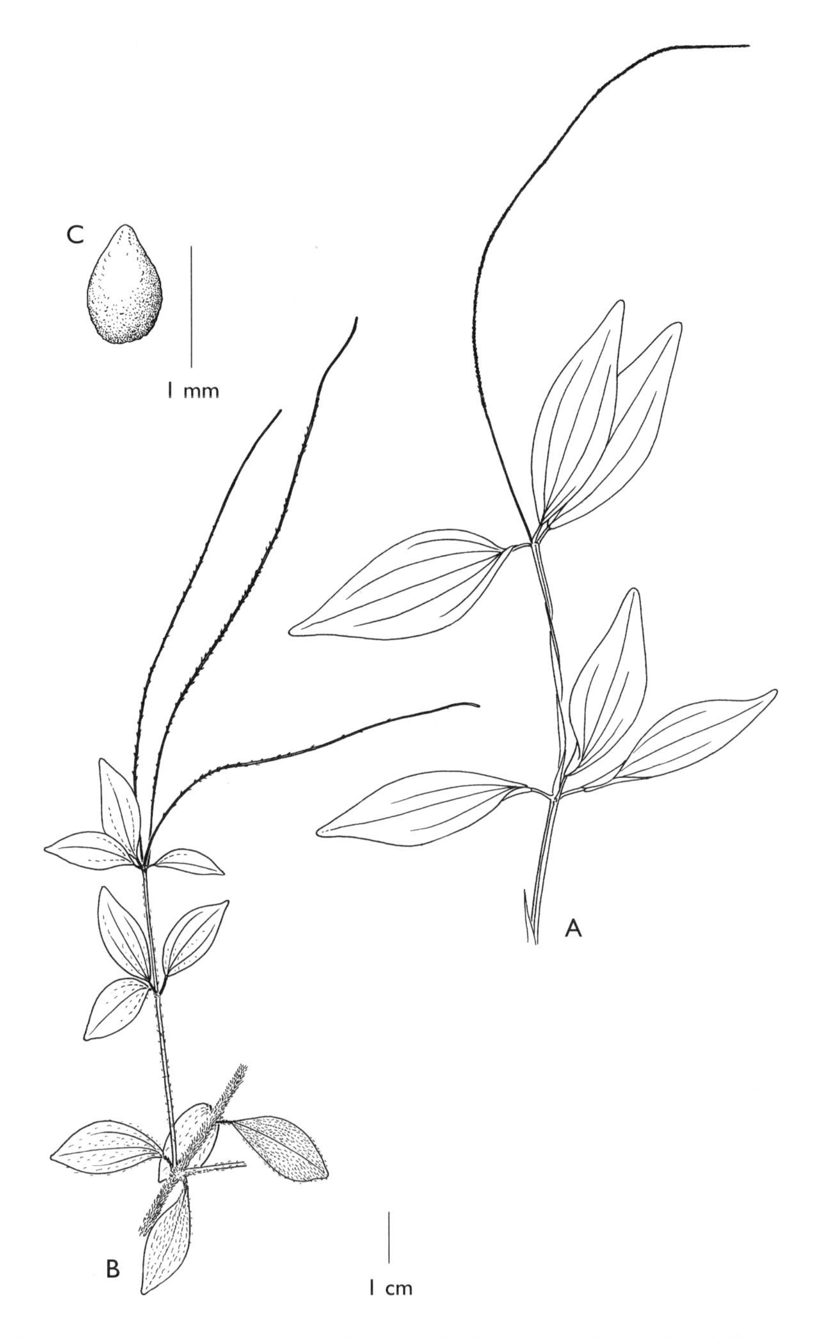

Fig. 5. *Peperomia angustata* Kunth: A, habit. *Peperomia blanda* (Jacq.) Kunth: B, habit; C, fruit (A, Feuillet 1173; B-C, Henkel 5658).

Distribution: Mexico, West Indies, Venezuela, Guyana, Suriname, south to Bolivia, Argentine, Peru; also in Australia; above 200 m, up to 1100 m elev.; 6 collections studied (GU: 5; SU: 1).

Specimens examined: Guyana: Cuyuni-Mazaruni region, Pipoly 10288 (U); Kanuku Mts., Wabuwak, Wilson-Browne 181 (BRG, K); Potaro-Siparuni region, Henkel 5658 (U); Upper Takutu – Upper Essequibo region, Chodikar R., Clarke 2809 (U); Rupununi Distr., Shea Rock, Jansen-Jacobs *et al.* 4840 (BRG, U). Suriname: Wilhelmina Mts., Stahel 393 (= BW 7225) (U).

Notes: In the Willdenow herbarium nr 747 seems to be an isotype of *Piper blandum.* On the label we find the name "v.d. Schott" suggesting that he was the collector. According to Vegter (Index Herbar., Collectors 2(6): 852. 1986), 'Richard van der Schot" accompanied N.J. Jacquin to the West Indies; maybe also to Venezuela?
In the Paris herbarium the Humboldt & Bonpland s.n. collection from Venezuela, between Caracas and Guara bears an isotype label. This later collection can not represent the type.

4. **Peperomia delascioi** Steyerm., Brittonia 40: 294. 1988. Type: Venezuela, Bolivar, Kamarkaibaray-tepui, Delascio 13161 (holotype VEN, isotype MO not seen). – Fig. 6 A-D

Terrestrial herb, non-creeping, suberect or ascending up to 9 cm, not rooting at nodes. Stem minutely hirtellous (spreading hairs 0.1 mm long); upper internodes 4-10 mm long. Uppermost leaves opposite, the others alternate; petiole 0.3-0.8 cm long, minutely hirtellous; blade basely attached or slightly peltate, fleshy-membranous, oblately suborbicular, 0.5-0.8 x 0.5-1 cm, margin not ciliate, apex emarginate with some hairs at notch, base rounded or subtruncate, glabrous above, minutely hirtellous at base below; palmately 3-5-veined. Inflorescence terminal opposite the uppermost leaves, erect, solitary; peduncle 0.3-0.5(-2) cm long, minutely hirtellous, without bracts; spike with scattered flowers, 0.6-3 cm long; floral bracts rounded, umbonate with pale margin, 0.3 mm in diam.; rachis glabrous. Fruits stipitate pale brown, smooth, oblong-ellipsoid, 1-1.1 x 0.5 mm, crowned with a persistent, conic style, stigma apical, mammiform.

Distribution: Known only from altitudes above 2000 m at the Guyana – Venezuela border; 1 collection studied of Guyana: Mazaruni-Potaro region, Roraima summit, La Proa camp near lake Gladys, Liesner 23288 (U) (GU: 1).

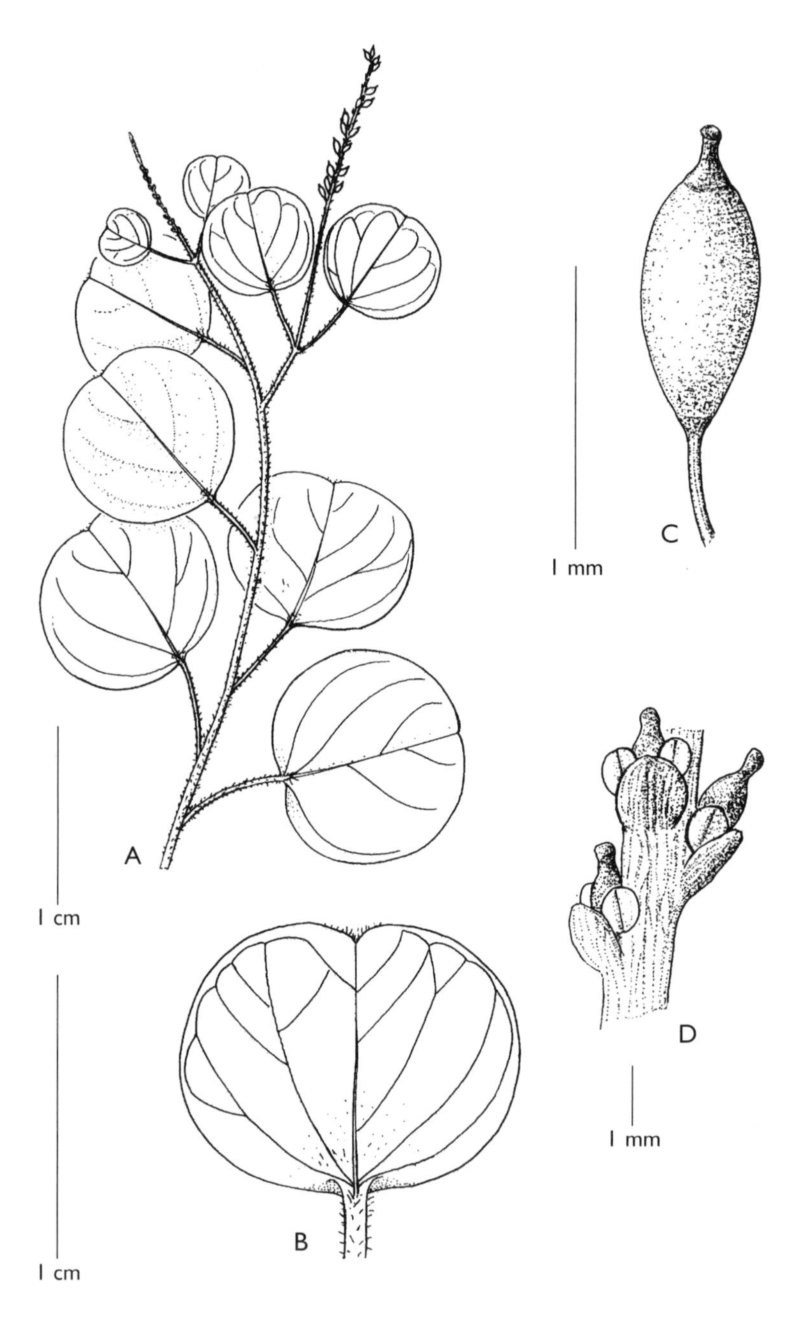

Fig. 6. *Peperomia delascioi* Steyerm.: A, habit; B, leaf; C, stipitate fruit; D, part of rachis with fruits. After Steyermark, Brittonia 40: 295. 1988.

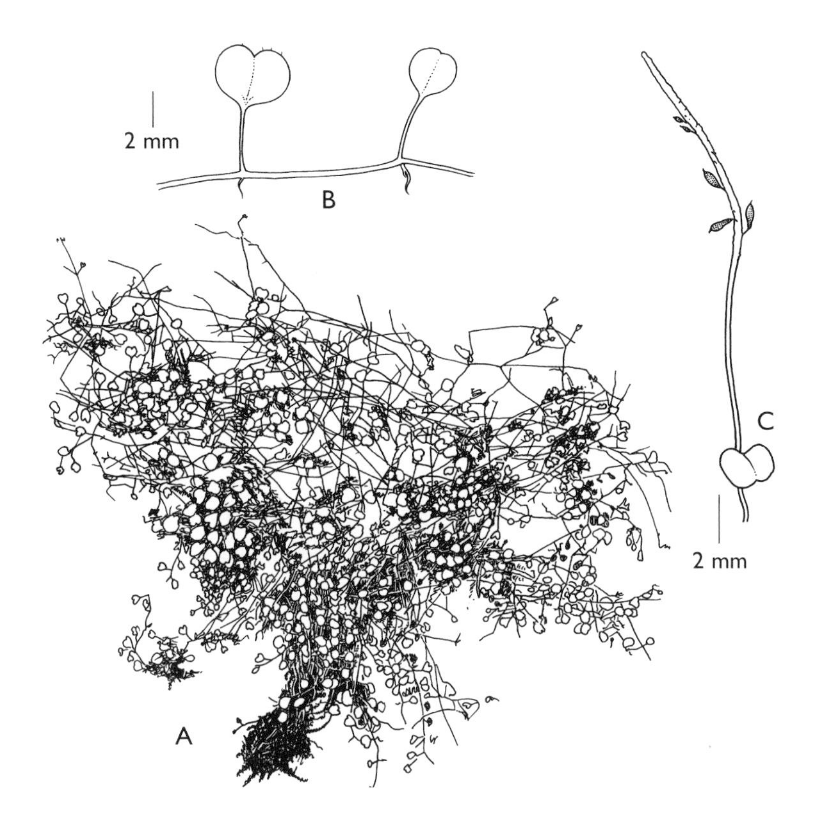

Fig. 7. *Peperomia emarginella* (Sw. ex Wikstr.) C. DC.: A, habit; B, part of stem with leaves; C, spike (A-B, de Granville *et al.* 8750; C, Clarke 9725).

Note: According to Steyermark, *Peperomia delascioi* can be distinguished from *P. emarginella* which also has an emarginate leaf apex and orbicular leaves by the following characters: *Peperomia delascioi* is suberect to ascending, has hirtellous stems, petioles and peduncles and non-ciliate leaves. From *P. rotundifolia*, it can be distinguished by the stipitate fruit with mammiform apex and by the non-creeping, not rooting habit. *Peperomia delascioi* is known only from cool summits in dwarf shrub and herbaceous vegetation at altitudes from 2400-2800 m, whereas *P. rotundifolia* prefers warm, humid forest from near sea level to 1000 m.

5. **Peperomia emarginella** (Sw. ex Wikstr.) C. DC. in A. DC., Prodr. 16(1): 437. 1869. – *Piper emarginellum* Sw. ex Wikstr., Kongl. Vetensk. Acad. Handl. 1827: 56. 1827. Type: Guadeloupe, Forsström s.n. (holotype S, not seen). – Fig. 7 A-C

Epiphytic or epilithic, or epiphyllous herb, creeping and rooting at nodes, glabrous or sparsely villous. Leaves alternate, basely attached; petiole 0.1-0.5 cm long, often red, glabrous or with sparse hairs; blade slightly fleshy, bright green, orbicular or subreniform, 0.15-0.6 cm in diam., margin somewhat ciliate, apex emarginate, base rounded or truncate, glabrous or with a few hairs; obsoletely palmately 3-veined. Inflorescence terminal, single, erect; peduncle slender, to 1 cm long, pinkish; spike to 1.5 cm long, laxly flowered; floral bracts sub-orbicular to oblong. Fruits basely attached (according to Steyermark, on a short stipe), ellipsoid, ca. 1 mm long with apical stigma.

Distribution: West Indies, C and S America, S to Brazil and Peru; from sea level to 600 m elev.; 12 collections studied (GU: 8; SU: 1; FG: 3).

Selected specimens: Guyana: Puruni R., Jenman 7597 (BRG, K); Essequibo R., Moraballi Cr., Sandwith 27 (K); Upper Takutu – Upper Essequibo region, Acarai Mts., Henkel 4964 (U). French Guiana: Mt. Galbao, de Granville *et al.* 8750 (CAY, NY, P, U, US); Saül, Boom 10858 (NY, U).

Vernacular name: Suriname: piki kopali (Boni; Sauvain 369).

Note: see note to *Peperomia delascioi.*

6. **Peperomia galioides** Kunth in Humb., Bonpl. & Kunth, Nov. Gen. Sp. ed. qu. 1: 71, t. 17. 1816. – *Piper galioides* (Kunth) Poir. in Lam., Encycl. Suppl. 4: 470. 1816. Type: Colombia, near Tequendama Falls, Humboldt & Bonpland s.n. (isotype B-W 762). – Fig. 8 A-B

Terrestrial, epilithic or epiphytic, cespitose herb. Stem minutely pubescent to glabrous, 10-80 cm long, green or green with red spots or reddish. Leaves in whorls of 1-4(-9), basely attached; petiole to 0.3 cm long, minutely pubescent; blade fleshy to subcoriaceous, narrowly spathulate, narrowly elliptic or narrowly oblanceolate, larger towards apex of stem, 0.5-3.0 x 0.2-0.6 cm, ciliate towards obtuse apex, base acutish, glabrous; venation slightly obvious, palmately 3-veined. Inflorescence terminal or axillary, often in groups of 2-5; peduncle not slender, 0.2-1.2 cm long, minutely pubescent to glabrous, winged, green; spike up to 15 cm long, pale yellow to green, laxly flowered; floral bracts rounded, glandular. Fruits subbasely attached, globose, brown, apex oblique with subapical stigma.

Fig. 8. *Peperomia galioides* Kunth: A, habit; B, fruit (A-B, Hahn 5019).

Distribution: West Indies, Mexico, C and S America, south to Brazil and Peru; 2 collections seen from Venezuela: Tachira, elev. 2250-2500 m, Hahn 5019 (U) and Bolivar, Rondon Camp, Mt. Roraima, 2700 m elev., Tate 458 (NY). It is expected to occur on Guyanan side of Mt. Roraima as well.

7. **Peperomia glabella** (Sw.) A. Dietr., Sp. Pl. 1: 156. 1831. – *Piper glabellum* Sw., Prodr. 16. 1788. Type: Jamaica, Swartz s.n. (holotype G-DC, not seen). – Fig. 9 A-C

Micropiper melanostigma Miq., Comm. Phytogr. 51. 1840. – *Peperomia melanostigma* (Miq.) Miq., Syst. Piperac. 90. 1843. Type: Suriname, Focke s.n. (holotype U).
Peperomia caulibarbis Miq., Syst. Piperac. 98. 1843. Type: Brazil, Ilha de Santa Catharina, Gaudichaud 287 (holotype G-DEL, isotype G-DC, none seen).
Peperomia velloziana Miq. var. *polysticta* Miq., Linnaea 18: 226. 1845. Type: Suriname, near plantation Tourtonne, Kegel 437 (isotype P).
Peperomia melanostigma (Miq.) Miq. var. *nervulosa* C. DC. in A. DC., Prodr. 16(1): 409. 1869. – *Peperomia glabella* (Sw.) A. Dietr. var. *nervulosa* (C. DC.) Yunck., Ann. Missouri Bot. Gard. 37: 98. 1950. Type: Suriname, Paramaribo, Hostmann 437 (holotype B not seen, isotypes G, K, BM, P, U).

Epiphytic, stoloniferous herb, strongly glandular dotted, erect, hanging, prostrate or ascending. Stem to 15 cm long, green or reddish, glabrous, except for lines of hairs decurrent from nodes. Leaves basely attached, alternate; petiole 0.5-1(-2.5) cm long, usually ciliate with 2 rows of hairs; blade fleshy to membranous, densely black-glandular (when dried), narrowly ovate or ovate, 1.5-8(-11) x 0.5-3.5(-5) cm, margin ciliate near apex, apex acute to acuminate, base acute to acuminate, sometimes subobtuse, glabrous; venation obvious, pli- or palmately 3-5-veined. Inflorescence terminal, single, or several in upper leaf axils, erect to pendent ; peduncle 0.5-1.5 cm long, glabrous, black-glandular; spike to 16 cm long, moderately densely flowered, flowers distinctly separate; floral bracts rounded, glabrous and black-glandular. Fruits laterally attached, globose, reddish brown, apex oblique with subapical stigma.

Distribution: West Indies, C and S America; from sea level up to 2400 m elev.; over 180 collections studied (GU: 33; SU: 43; FG: 112).

Selected specimens: Guyana: upper Essequibo R., Rewa R., near Corona Falls, Jansen-Jacobs *et al.* 5854 (BRG, U); Barima-Waini region, along Barima R., McDowell 4397 (U, US). Suriname: Kabalebo area, Lindeman & de Roon *et al.* 406 (BBS, NY, U); Tumac Humac Mts.,

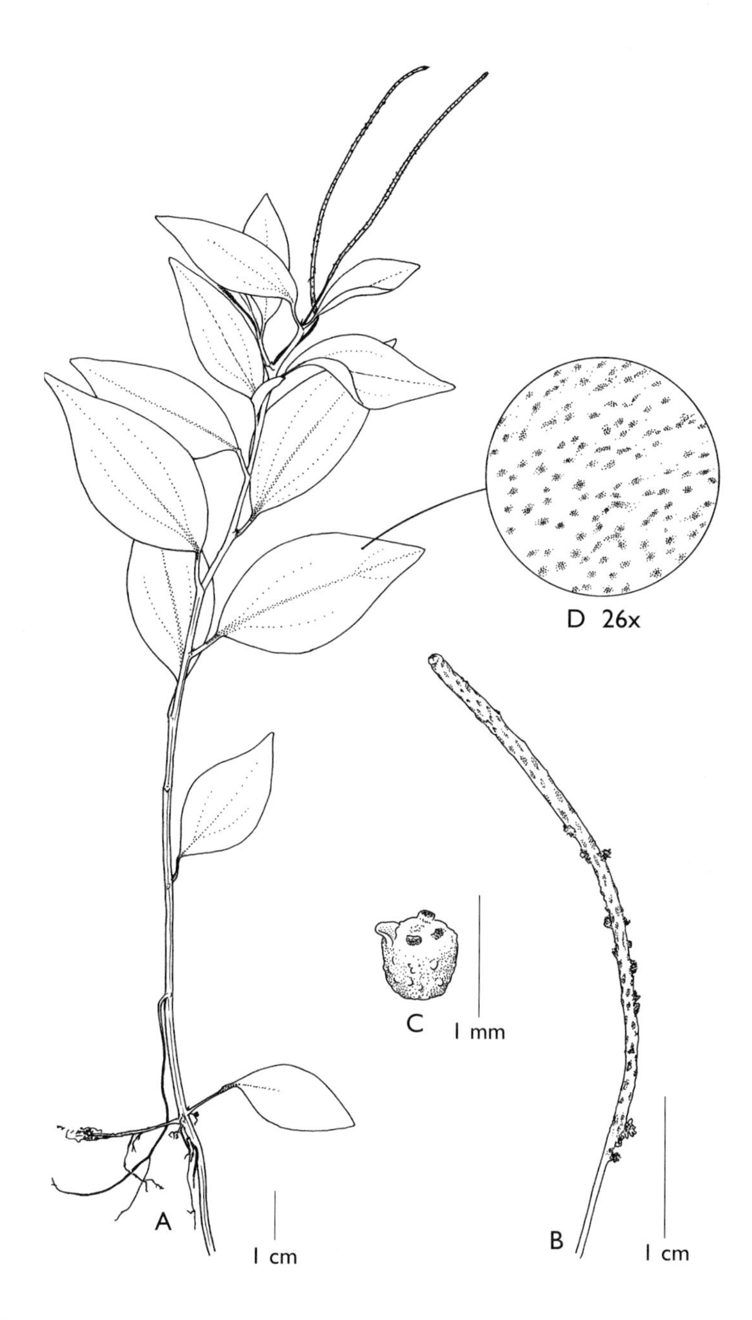

Fig. 9. *Peperomia glabella* (Sw.) A. Dietr.: A, habit, with detail of glandular leaf surface; B, spike with fruits; C, fruit with 3 glands and subapical stigma (A, Kegel 437; B-C, van Donselaar 3819).

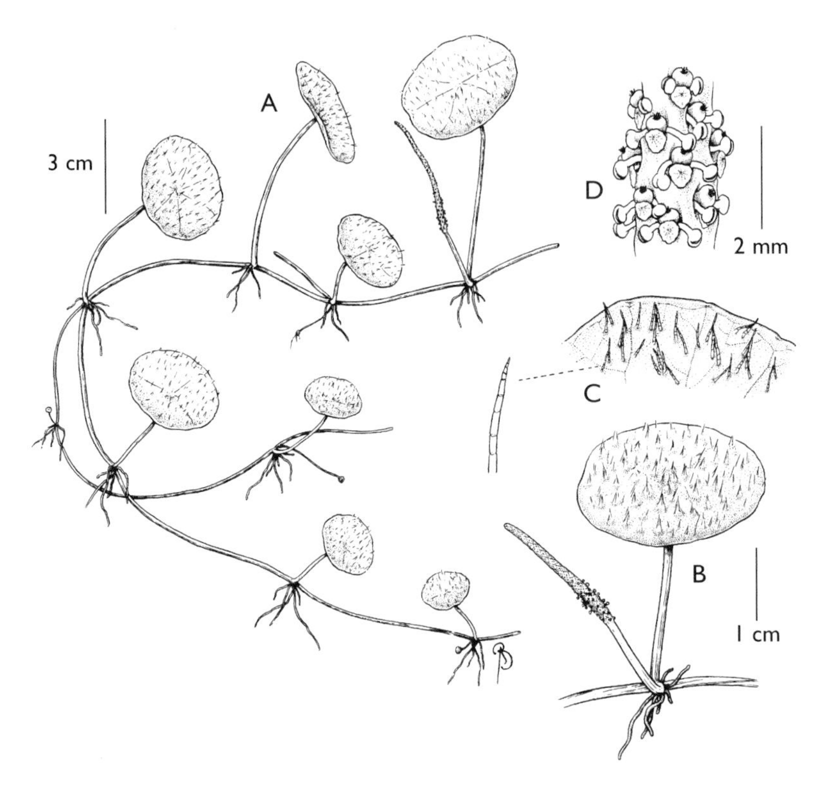

Fig. 10. *Peperomia gracieana* Görts: A, habit; B, part of stem with leaf and spike; C, detail of leaf surface and trichoom; D, part of flowering spike showing floral bracts and flowers with stamens (A, Mori *et al.* 23264; B-D, Mori *et al.* 23773). Drawing by B. Angell (NY).

Acevedo 5947 (U, US). French Guiana: Upper Marouini R. basin, Roche Koutou, de Granville 9283 (CAY, NY, P, U, US); Saül, Maas *et al.* 2236 (CAY, K, NY, U).

Phenology: Flowering throughout the year.

Notes: In the Paris herbarium there is a cover containing 3 isotypes, 2 of which are evidently Hostmann 437 – one of them also mentioning the name Kappler – the third voucher also with collection number 437 bears the name Kegel. It was used to describe *Peperomia velloziana* var. *polysticta*. As a result, it can be concluded that the latter is conspecific with *P. glabella*.

Trelease & Yuncker, l.c.: 587-590 and Steyermark, l.c.: 109 separate the Guianan taxa into 2 resp. 3 varieties. The main differences are the size of the leaves. In a very variable species it seems unnecessary to divide the taxon.
P. glabella is easily recognizable (when dried) by the multitude of dark glandular dots on most parts.

8. **Peperomia gracieana** Görts, Brittonia 50: 56. 1998. Type: French Guiana: near Saül, E of Eaux Claires, Mori *et al.* 23773 (holotype NY, isotypes CAY, U). – Fig. 10 A-D

Epiphytic, terrestrial or epilithic herb, creeping and rooting at nodes. Stem glabrous, 40 cm long, whitish green. Leaves alternate, distinctly peltate; petiole 1-5(-10) cm long, glabrous; blade fleshy *in vivo*, membranous almost transparent when dried, broadly elliptic to orbicular, 1-4.5 x 1-3.5 cm, apex rounded, base rounded, pilose with erect trichomes on upper surface; venation obvious in dried state, veins 6-8 from base. Inflorescence single, axillary; peduncle slender, 1.5-1.7 cm long, glabrous, green; spike laxly flowered, 4-6 cm long, green; floral bracts rounded, glabrous. Young fruits flat at apex, green, mature fruits not yet collected.

Distribution: Only known from French Guiana and Brazil (Amapa); in non-flooded moist forest, creeping and forming loose mats along creeks on rocks, on the forest floor, or on fallen logs, between 150-500 m alt; 17 collections studied (FG: 17).

Selected specimens: French Guiana: Vicinity of Saül: ca. 10 km NW from Eaux Claires, Mori *et al.* 23264 (NY); Massif des Emerillons, N slope, Cremers 6696 (CAY, K, NY, P, U, US); Haut Tampoc, Cremers 4576 (CAY?, U); Crique Cacao - Bassin de la Haute Camopi, Transect I, Prévost & Sabatier 2425 (CAY, U); Saül, Görts *et al.* 76 (P, U).

9. **Peperomia haematolepis** Trel. in J.F. Macbr., Field Mus. Nat. Hist., Bot. Ser. 13(2): 52. 1936. Type: Peru, Junín, hacienda Chalhuapuquio, Stevens 212 (holotype ILL, not seen). – Fig. 11 A

Epiphytic, glabrous herb. Stem ascending. Leaves alternate; petiole not grooved or winged, 0.5-1.5 cm long; blade ellipic to obovate (to broadly ovate in extra-Guianan specimens), 4.5 x 3-4 cm, apex rounded or emarginate, base cuneate; pinnately veined veins branching off in lower part of primary vein, may be obsolete when dried. Inflorescence

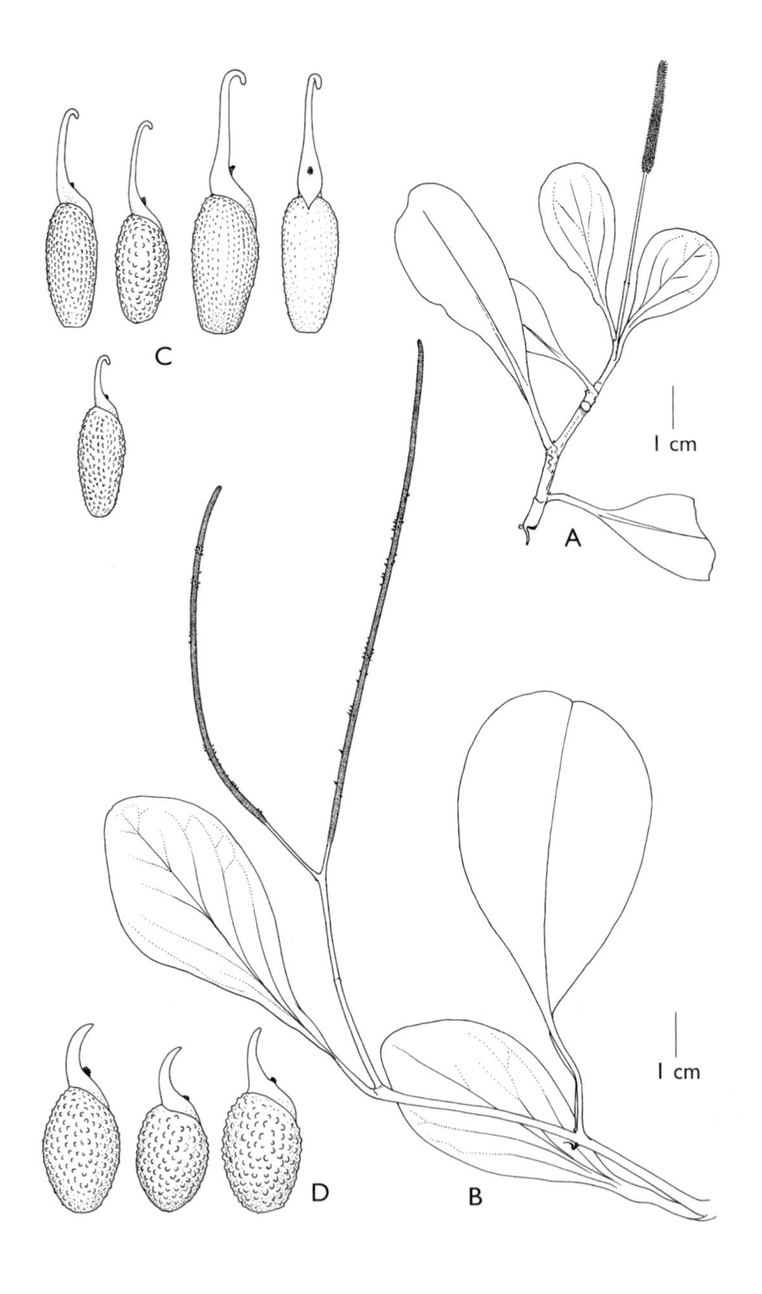

Fig. 11. *Peperomia haematolepis* Trel.: A, habit. *Peperomia magnoliifolia* (Jacq.) A. Dietr.: B, habit; D, fruits. *Peperomia obtusifolia* (L.) A. Dietr.: C, fruits (A, Görts-van Rijn *et al*. 71; B, Chanderbali *et al*. 603). C-D from Table 2 in: H. Dahlstedt, Kongl. Svenska Vetensk. Handl. Stockholm 33(2): Tafel 2. 1900.

terminal, single, erect; peduncle not slender, 3.5-5 cm long, glabrous or minutely puberulent; spike 2-5 cm long, reddish; floral bracts rounded, glandular. Fruits ellipsoid or ovoid narrowing into slender beak, with curved tip, stigma at base of beak.

Distribution: Colombia, Venezuela, Suriname, French Guiana, Ecuador and Peru, 200-700 m elev.; 8 collections studied (SU: 1; FG: 2).

Specimens examined: Suriname: Lely Mts., Lindeman & Stoffers *et al.* 417 (U). French Guiana: Saül, Görts-van Rijn & Gouda 71 (CAY, BY, U); Mt. de l'Inini, Feuillet 3812 (CAY, U).

Notes: *Peperomia haematolepis* is rather similar to small sized specimens of *P. obtusifolia* or *P. magnoliifolia* but differs in having shorter spikes (to 5 cm vs. to 18 cm), and peduncle shorter or not much longer than the spike (vs. spike much longer than peduncle in the latter two species).
Some of the collections that have been identified as *P. obtusifolia* could belong to *P. haematolepis.*

10. **Peperomia hernandiifolia** (Vahl) A. Dietr., Sp. Pl. 1: 157. 1831, as 'hernandiaefolia'. – *Piper hernandiifolium* Vahl, Enum. Pl. 1: 344. 1804, as 'hernandifolium'. Type locality: West Indies; type not designated. – Fig. 12 A-B

Peperomia choroniana C. DC. var. *puberulenta* Yunck. in Trel. & Yunck., Piperac. N. South Amer. 730. 1950. Type: Venezuela, Amazonas, Cerro Duida, Steyermark 57991 (holotype F, not seen).

Epiphytic, terrestrial or epilithic herb, creeping. Stem to 0.4 cm thick when dried, retrorsely minutely crisp-pubescent, green and reddish. Leaves alternate, distinctly peltate; petiole attached 1-3 cm from base, 3-11(-16) cm long, minutely pubescent, glabrescent; blade coriaceous or subcoriaceous, narrowly ovate, ovate or almost orbicular, 4.5-8(-12) x 3-5.5(-8) cm, margin somewhat ciliate, apex acuminate, base rounded, minutely pubescent, glabrescent; veins 9-11 arising from base. Inflorescence terminal or axillary, solitary; peduncle not slender, 1.5-5.5 cm long, minutely pubescent, green; spike 3-5 cm long, densely flowered; floral bracts rounded, glabrous, glandular. Fruits basely attached, ellipsoid, reddish brown, apex with slender beak and stigma at base of beak.

Distribution: West Indies, Mexico, C America and northern S America into Brazil; elev. 500-2000 m; 6 collections studied (GU: 6).

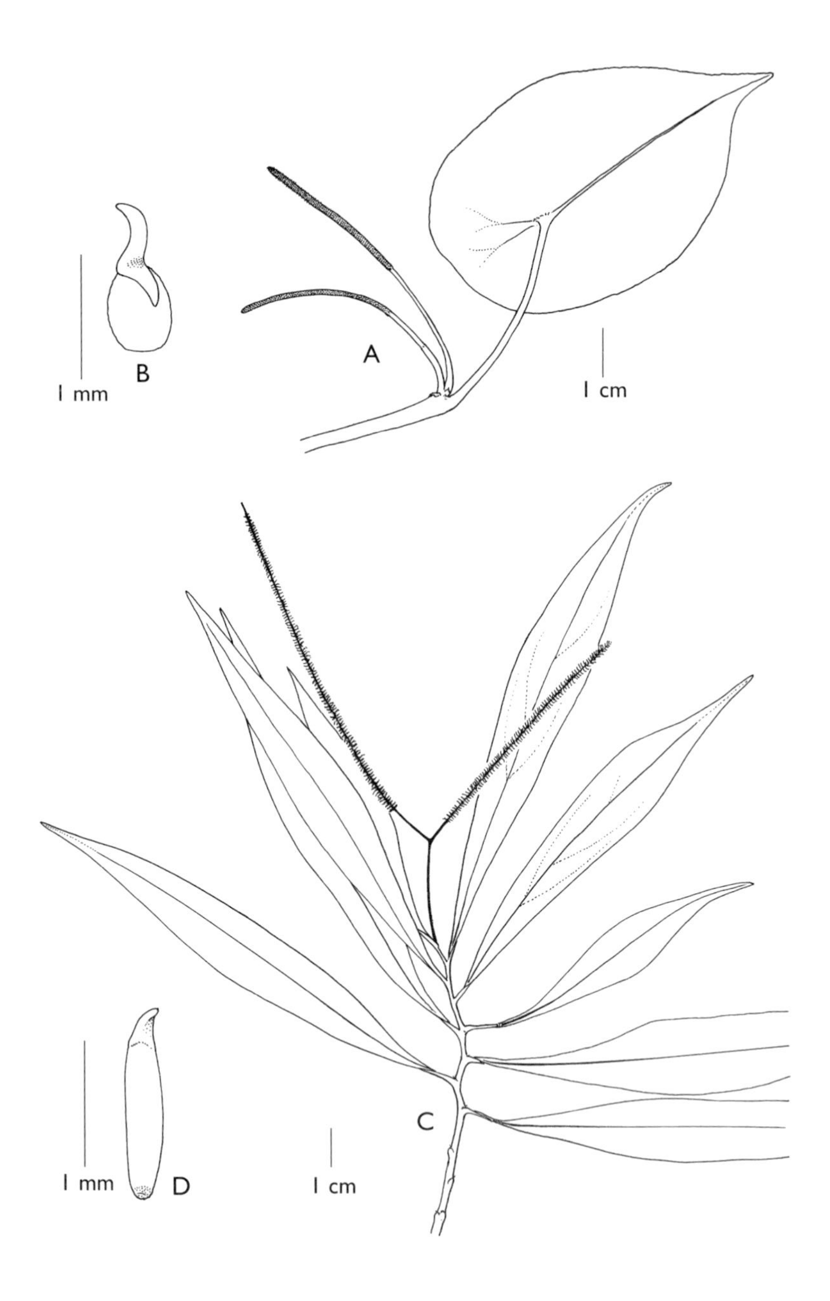

Fig. 12. *Peperomia hernandiifolia* (Vahl) A. Dietr.: A, habit; B, fruit. *Peperomia lancifolia* Hook. subsp. *lancifolia*: C, habit; D, fruit (A-B, Maas *et al.* 4442; C-D, Henkel 4419).

Specimens examined: Guyana, Pakaraima Mts.: Mt. Kovak, Maas & Westra 4442 (K, U); Ireng R., Mutchnick 257 (U); Mt. Roraima, Paikwa Trail, R. Persaud 53 (K); Cuyuni-Mazaruni region, NW of N prow Mt. Roraima, Hahn 5456 (U, US), Renz 14199 (U); Paruima, Clarke 5592 (U).

11. **Peperomia lancifolia** Hook. , Icon. Pl. 4: ad t. 332. 1840. Type: Mexico, Jalapa, Galeotti s.n. (holotype K, not seen). – Fig. 12 C-D

In the Guianas only: subsp. **lancifolia**.

Terrestrial, epilithic or epiphytic herb, erect or ascending and rooting at decumbent base. Stem glabrous, 10-34 cm long, green or reddish. Leaves basely attached, alternate; petiole 0.5-1 cm long, glabrous, narrowly winged; blade fleshy and membranous, narrowly elliptic or narrowly oblanceolate, (3.5-)7-17 x (0.7-)1-3 cm, somewhat ciliate towards apex, apex acuminate, base acute, decurrent into petiole, glabrous; pinnately 7-11-veined. Inflorescence terminal or in upper leaf axils, solitary, bifurcate or once-branched; common peduncle green, not slender, 2-6 cm long, lateral branches 0.5-1.3 cm long, with linear-(ob)lanceolate, cadoucous bracts; spike densely flowered, erect, 4-11 cm long, whitish or yellowish green; floral bracts rounded, glabrous, glandular. Fruits basely attached, narrowly cylindrical, brown, apex with short beak and subapical stigma.

Distribution: Mexico, C America, and from S Venezuela and Guyana to W Brazil and Peru; from 700-1600 m elev.; 12 collections studied (GU: 12).

Selected specimens: Guyana: Potaro-Siparuni region, Pakaraima Mts., Henkel 1405 (U); Mt. Kopinang, Hahn 4314 (U); escarpment to foot of Kopinang Falls, Maguire *et al.* 46076A (NY); Cuyuni-Mazaruni region, Mt. Ayanganna, Pipoly 11083 (U); Pakaraima Mts., Mt. Wokomong, Henkel 4455 (U); Upper Mazaruni Basin, Tillett *et al.* 44973 (NY, U).

Notes: In the non-Guyanan *Peperomia lancifolia* Hook. subsp. *erasmiiformis* (Trel.) Steyerm., the inflorescence is paniculate.
The collection Maguire 46076A had been misidentified as *P. pyramidata* Sodiro, the specimen fits well within *P. lancifolia* having a once-branched panicle, not a compound one as in *P. pyramidata* from Ecuador.

12. **Peperomia macrostachya** (Vahl) A. Dietr., Sp. Pl. 1: 149. 1831. – *Piper macrostachyon* Vahl, Enum. Pl. 1: 341. 1804. Type: French Guiana, Cayenne, L.C. Richard s.n. (holotype P, not found). – Fig. 13 A-G

Fig. 13. *Peperomia macrostachya* (Vahl) A. Dietr.: A, habit; B-E, leaf shapes; F-G, fruits (A, F, Gillespie 1180; B, Acevedo 3415; D, G, de Granville *et al.* 9560; C, E, Mutchnick 1102).

Piper myosuroides Rudge, Pl. Guian. Rar. 9, t. 5. 1805. – *Peperomia myosuroides* (Rudge) A. Dietr., Sp. Pl. 1: 157. 1831, as 'myusoroides'. Type: French Guiana, J. Martin s.n. (holotype BM).
Peperomia elongata Kunth in Humb., Bonpl. & Kunth, Nov. Gen. Sp. ed. qu. 1: 62. 1816. Type: Venezuela, Edo. Monagas, Caripe, Humboldt & Bonpland s.n. (holotype P, isotypes B-W 717, G).
Peperomia nematostachya Link, Jahrb. Gewächsk. 1(3): 63. 1820. – *Peperomia macrostachya* (Vahl) A. Dietr. var. *nematostachya* (Link) Trel. & Yunck., Piperac. N. South Amer. 661. 1950. Type: Brazil, von Hoffmannsegg s.n. (holotype B-W 720, isotype G).
Peperomia myriocarpa Miq., Syst. Piperac. 185. 1843. Type: Brazil, Minas Gerais, Claussen s.n. (holotype G-DC, not seen).
Peperomia parkeriana Miq., London J. Bot. 4: 427. 1845. Type: Guyana, Parker s.n. (holotype K).
Peperomia piperea C. DC., J. Bot. 4: 143. 1866. Type: Guyana, Parker s.n. (holotype K).
Peperomia surinamensis C. DC. in A. DC., Prodr. 16(1): 408. 1869. Type: Suriname, near Paramaribo, Kappler 1577 (lectotype P, isolectotype U, photo of G isolectotype NY) (here designated).
Peperomia elongata Kunth var. *guianensis* Yunck., in Maguire *et al.*, Bull. Torrey Bot. Club 75: 290. 1948. Type: Guyana, Pomeroon Distr., Kamwatta, de la Cruz 1197 (holotype US, isotype NY).
Peperomia elongata Kunth var. *piliramea* Trel. & Yunck., Piperac. N. South Amer. 659. 1950. Type: Guyana, near mouth of Onoro R., A.C. Smith 2650 (holotype US, isotypes F, MO, U).

Epiphytic herb, often growing on ants' nests, creeping or hanging and rooting at nodes. Stem glabrous, to 100 cm long, green or green with red nodes. Leaves alternate, basely attached; petiole 0.5-2 cm long, glabrous, ciliate or puberulent, reddish; blade fleshy-coriaceous or subcoriaceous, narrowly elliptic to ovate, 4-11 x 1.5-6 cm, more or less ciliate, apex acute or acuminate, base rounded or cordulate or acutish, glabrous or slightly pubescent, a waxy exudate like small scales or stars is usually present on upper surface; venation obsolete, 7-9-pli-veined. Inflorescence terminal, pendent, few together; peduncle not slender, to 1.5(-3) cm long, glabrous or minutely pubescent or papillose, red or brown; spike to 30 cm long, yellow to green, densely flowered; floral bracts rounded, dentate or fimbriate, glandular. Fruits basely attached, cylindrical, verruculose, brown, apex obliquely scutellate with subapical stigma.

Distribution: Mexico, C and northern S America, to Brazil and Peru; in wet forest, from sea level to 1800 m elev.; over 250 collections studied (GU: 103; SU: 52; FG: 105).

Selected specimens: Guyana: Essequibo R., Tiger Cr., Sandwith 1180 (K, NY); Marudi Mts., Stoffers & Görts-van Rijn *et al.* 246 (BRG, K, NY, U). Suriname: Oelemari R., Wessels Boer 987 (K, NY, U);

Coronie Distr., Jenny-Coronie, LBB 15130 (BBS). French Guiana: Kapiri Cr., Basin de l'Approuague, Cremers *et al.* 11529 (CAY, NY, P, U, US); Roche Forteresse, Roche Koutou, de Granville *et al.* 9560 (CAY, NY, P, U, US).

Vernacular names: Guyana: kutumi, kararafaru, kararafarumafartaino (Guppy 434).

Note: The differences between *P. elongata* and its varieties and *P. macrostachya* are in my opinion too little to separate the species. I agree with Burger (1971: 49), who said that they all belong to one, diverse, wide ranging species (complex).

13. **Peperomia magnoliifolia** (Jacq.) A. Dietr., Sp. Pl. 1: 153. 1831, as 'magnoliaefolia'. – *Piper magnoliifolium* Jacq., Collecteana 3: 210. 1791, as 'magnoliaefolia'. Type: not designated. – Fig. 11 B, D

Epiphytic, epilithic or terrestrial herb. Stem erect, creeping or ascending, rooting at nodes, 10-50 cm long, glabrous, green or green with red spots. Leaves alternate, basely attached; petiole 1-7 cm long, more or less laterally winged or grooved, glabrous; blade fleshy, often red-dotted, obovate or elliptic, sometimes even broadly elliptic, 4.5-14 x 2.5-6 cm, apex rounded, obtuse occasionally slightly emarginate to acutish, base cuneately decurrent into petiole, sometimes abruptly contracted, glabrous; venation obvious, ca. 7-pli-veined, or a few veins branching off primary vein or obsolete when dried. Inflorescence terminal, erect, solitary or usually 2(-few) together; peduncle slender (when dried) or not slender, 1-8 cm long, glabrous, green or pinkish; spike to 18 cm long, green, white or yellow, densely flowered; floral bracts rounded, 5-7 mm in diam., glabrous, glandular. Fruits basely attached, subglobose, verruculose, reddish brown or blackish, gradually tapering into beak, which is slightly hooked at the very tip, stigma at base of beak.

Distribution: West Indies, from C to northern S America, to Brazil and Colombia; also in Micronesia; from sea level to over 1200 m elev.; over 40 collections studied (GU: 15; SU: 13; FG: 14).

Selected specimens: Guyana: Mabura hill, Stoffers & Görts-van Rijn *et al.* 116 (K, NY, U); Upper Takatu - Upper Essequibo region, Rewa R., near Great Falls, Clarke 6618 (BRG, U, US). Suriname: Wilhelmina Mts., Maguire *et al.* 54012 (K, NY, U), 2 km above confluent Lucie R., Irwin *et al.* 55784 (K, NY, U). French Guiana: Mt. Chauve, Cremers *et al.* 15235 (CAY, NY, P, U, US); Roche Touatou, Oyapock basin, Cremers *et al.* 14121 (CAY, NY, P, U, US).

Notes: For differences with *Peperomia obtusifolia*, see note to the latter. Especially dried specimens of *P. magnoliifolia* have a strong smell of lovage (*Levistictum officinale*, Apiaceae).

14. **Peperomia maguirei** Yunck. in Maguire *et al.*, Bull. Torrey Bot. Club 75: 291. 1948. Type: Suriname, Tafelberg, near SE margin Arrowhead Basin, Maguire 24420 (holotype NY, isotypes ILL, U, US).
– Fig. 14 A-B

Epiphytic or epilithic herb, erect or ascending from prostrate, rooting base. Stem loosely villous, 10 cm long when fruiting, green or reddish brown. Leaves in whorls of 1-4, basely attached; petiole 0.2-0.6 cm long, loosely villous; blade fleshy coriaceous, sometimes pink below, obovate or somewhat elliptic, 1-3.5 x 0.6-2.3 cm, margin not ciliate, apex rounded to emarginate, base acute, glabrous or loosely villous on upper surface, glabrescent, glabrous below; veins 3, obvious. Inflorescence solitary, terminal, erect; peduncle 1-2 cm long, villous, green or reddish brown; spike 2-7 cm long, green, densely flowered, rachis hardly ridged; floral bracts rounded, with scariose margin, glabrous, glandular. Fruits not sunken, subbasely attached, globose, green, apex with short, thick style and apical stigma.

Distribution: The Guianas; up to 700 m elev.; 33 collections studied (GU: 4; SU: 17; FG: 12).

Selected specimens: Guyana: Mabura, ter Steege 43, 79 (U). Suriname: Lindeman & Stoffers *et al.* 474 (K, U); Wilhelmina Mts., Irwin *et al.* 54977 (NY, U). French Guiana: Approuague Basin, Nouragues, de Granville *et al.* 11087 (CAY, U); Tumac Humac Mts., Telouakem Inselberg, de Granville *et al.* 12317 (BBS, CAY, U, US).

Note: *Peperomia maguirei* has some resemblance to *P. quadrangularis* but differs in having mostly obovate leaves that are slightly emarginate (with hairs in the indentation) vs. usually broadly elliptic leaves with continuous apex. The leaves in *P. maguirei* are in whorls of 1-4, whereas in *P. quadrangularis* they are strictly opposite. The fruits of the latter are sunken in the rachis, which is distinctly ridged (at least when dried), whereas the fruits of *P. maguirei* are not sunken and the rachis is not obviously ridged.

15. **Peperomia obtusifolia** (L.) A. Dietr., Sp. Pl. 1: 154. 1831. – *Piper obtusifolium* L., Sp. Pl. 30. 1753. Type: [icon] Plumier, Descr. Pl. Amér. t. 70. 1693 (according to Howard 1973: 392). – Fig. 11 C

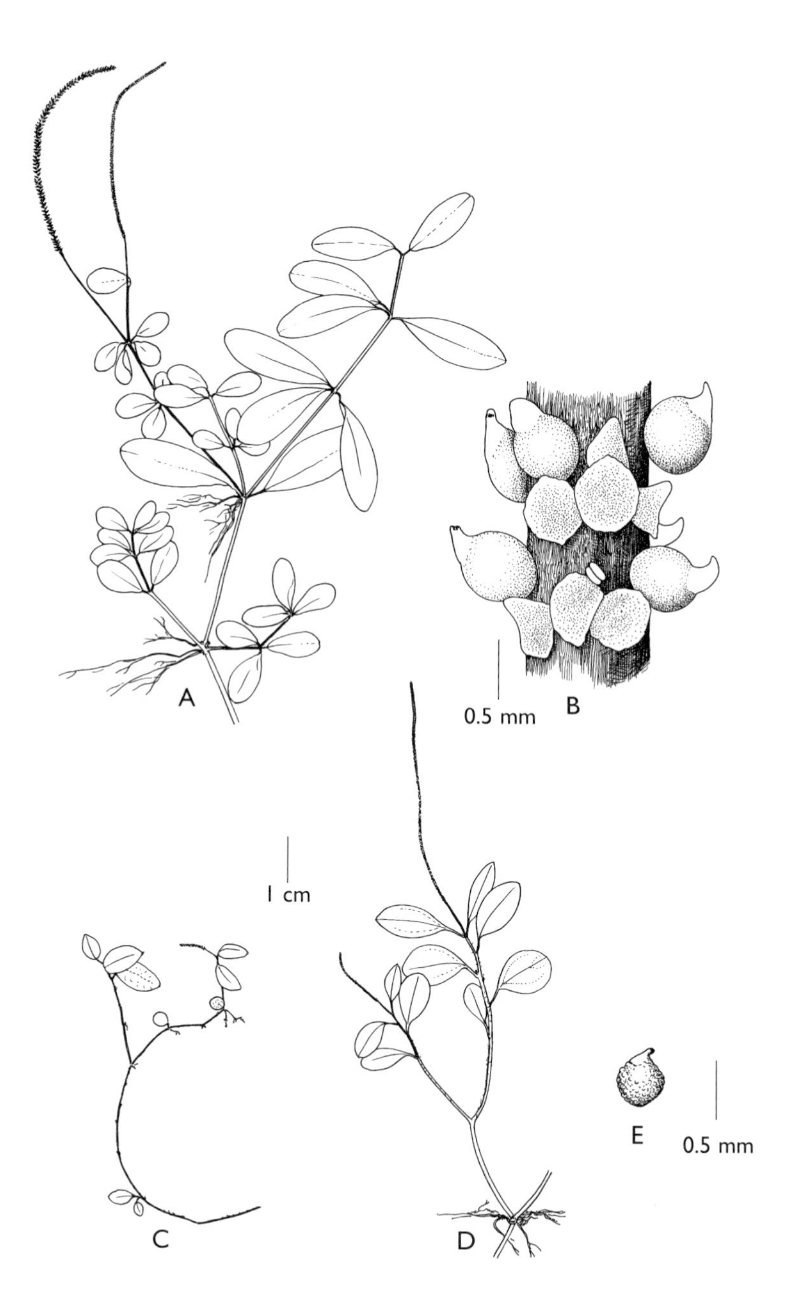

Fig. 14. *Peperomia maguirei* Yunck.: A, habit; B, part of spike with fruits. *Peperomia ouabianae* C. DC.: C, D, habit; E, fruit (A-B, de Granville *et al.* 11087; C, E, de Granville *et al.* 12316; D, Maas *et al.* 4448).

Piper cuneifolium Jacq., Collectanea 4: 127. 1791. – *Peperomia cuneifolia* (Jacq.) A. Dietr., Sp. Pl. 1: 146. 1831. Type locality: Venezuela, Caracas; type not designated.

Epiphytic, epilithic or terrestrial herb. Stem erect, creeping or ascending, rooting at nodes, glabrous, green or green with red spots. Leaves alternate, basely attached; petiole green or green with red spots, 1-6.5 cm long, more or less grooved, minutely pubescent; blade shiny green, pale green below, coriaceous, often with red spots, elliptic, obovate or subspathulate, 4-14.5 x 2.5-6.5 cm, apex rounded, obtuse or emarginate, base cuneate, glabrous; venation obvious or hardly visible, ca, 7-pli-veined, or few veins branching off primary vein. Inflorescence terminal, solitary or few together; both the common peduncle and those of individual spikes bracteate, green or pinkish, slender (when dried) or not slender, together 1-14 cm long, usually minutely pubescent; spikes to 18 cm long, greenish white to yellow, densely flowered; floral bracts rounded, 2-4 mm in diam., glabrous, glandular. Fruits basely attached, ellipsoid, verruculose, reddish brown or blackish, apex abruptly contracted into slender beak, which is curved or curled at tip, stigma at base of beak.

Distribution: Mexico, West Indies, C and northern S America; from sea level to over 1200 m elev.; ca. 100 collections studied (GU: 37; SU: 25; FG: 36).

Selected specimens: Guyana: NW Distr., Görts-van Rijn *et al.* s.n. (U); Barima-Waini region, McDowell 4475 (U). Suriname: Tafelberg, Maguire 24800 (K, U). French Guiana: Saül, Boom 10808 (NY, U); Route Cayenne - St. Laurent, Cremers *et al.* 5943 (CAY, P, U); Inini R., de Granville *et al.* 7464 (CAY, P, U).

Notes: There is quite some confusion about the limitation of *Peperomia magnoliifolia* and *P. obtusifolia*. Burger (1971: 52) considered the two taxa to be conspecific. Several authors, e.g. Howard (1973: 386) and Steyermark (1984: 165), have arguments to keep *P. obtusifolia* and *P. magnoliifolia* apart. Not only the morphological characters may distinguish the two taxa, like the indument of the petiole (minutely pubescent in *P. obtusifolia* vs. glabrous in *P. magnoliifolia*), the size of the bracts (diameter 2-4 mm in *P. obtusifolia* vs. 5-7 mm in *P. magnoliifolia*), and the shape of the fruit and the beak: (ellipsoid, abruptly contracted into the beak which is curved or curled at the tip in *P. obtusifolia* vs. gradually tapering into the beak, which is slightly hooked at the very tip in *P. magnoliifolia*). Also the chromosomal analyses show differences (information from José & Sharma). These arguments seem sound and it was not the right decision made by Kramer & Görts (1968: 420) to take the two species together.

After having studied the material at hand and with the knowledge of the morphology of the living plants G. Mathieu and I are convinced that there are two separate taxa.
It is difficult to tell sterile dried specimens apart. The living plants, however, may be recognized by leaf texture: *P. magnoliifolia* being fleshy, *P. obtusifolia* coriaceous.
In herbaria many specimens may have been incorrectly named. Both species do not seem to develop many fruits. On the long spikes often only a few can be found, sometimes even abortive. The beak, however, has also then been developed.
Steyermark, l.c.: 162, discerned the following four varieties: *P. obtusifolia* var. *obtusifolia,* var. *emarginulata* (C. DC.) Trel. & Yunck., var. *emarginata* (Ruiz & Pav.) Dahlst. and var. *cuneata* (Miq.) Griseb. They are difficult to separate. *P. obtusifolia* var. *emarginata* is known to occur in Venezuela (Tachira) from 450-650 m elev.; no collections have been reported from the Guianas, whereas var. *emarginulata* may be represented by a French Guianan collection from Kaw Mts. (Cremers 12427).
See also note to *P. haematolepis.*

16. **Peperomia ouabianae** C. DC., Candollea 1: 400, 410. 1923. – *Peperomia silvestris* C. DC., Notizbl. Bot. Gart. Berlin-Dahlem 7(62): 494. 1917, non *Peperomia sylvestris* C. DC. 1866. Type: Brazil, Mt. Roraima, near Ouabiana, Ule 8590 (holotype B, isotypes G, K, MG, fragment (DPU), NY). – Fig. 14 C-E

    Peperomia tafelbergensis Yunck. in Maguire *et. al.*, Bull. Torrey Bot. Club 75: 292. 1948. Type: Suriname, Tafelberg, Arrowhead Basin, Maguire 27170 (holotype NY, isotype U).

Epiphytic or epilithic herb, creeping and rooting at nodes. Stem loosely villous, 3-8 cm long, green to reddish. Leaves alternate; petiole to 0.5 cm long, loosely villous; blade glandular-dotted, oblong-elliptic or oblong-obovate or almost orbicular, 0.5-2 cm x 0.3-0.9 cm, margin ciliate, apex obtuse, base acute to obtuse, sparsely to densely villous above, less so beneath, or glabrescent; venation obvious, palmately 3-veined. Inflorescence terminal, solitary; peduncle slender, to 1.5 cm long, glabrous or minutely pubescent, green; spike 1.5-5.5 cm long, sparsely to densely flowered; floral bracts rounded, glabrous, glandular. Fruits sessile, ovoid, brown, somewhat verruculose, apex oblique with subapical stigma.

Distribution: Nicaragua, Costa Rica, Colombia, Venezuela, the Guianas and Brazil to Paraná and Rio de Janeiro; from sea level to 850-1400 m elev.; 64 collections studied (GU: 19; SU: 13; FG: 32).

Selected specimens: Guyana: Kanuku Mts., Jansen-Jacobs *et al.* 3571 (BRG, K, U); Potaro-Siparuni region, Pakaraima Mts., Clarke 1175 (BRG, U, US). Suriname: Brownsberg, Christenhusz 2642 (TUR, U); Lely Mts., Lindeman & Stoffers *et al.* 95 (BBS, U). French Guiana: Saül, Görts-van Rijn *et al.* 103 (CAY, NY, U); Massi, Inselberg Telouakem, Tumac Humac Mts., de Granville *et al.* 12316 (B, BBS, CAY, P, U, US).

Notes: In 1866, C. de Candolle had already described *Peperomia sylvestris* from Ecuador (J. Bot. 4: 142).
In Jörgensen, P.M. & S. León-Yánez, Catalogue of the Vascular Plants of Ecuador, 1999, *P. ouabianae* is given as being conspecific with *P. pilicaulis* C. DC. The species are rather similar but differ in indument, leaf shape and size of fruit. It seems better to keep the two species apart.

17. **Peperomia pellucida** (L.) Kunth in Humb., Bonpl. & Kunth, Nov. Gen. Sp. ed. qu. 1: 64. 1816. – *Piper pellucidum* L., Sp. Pl. 30. 1753. Type: [icon] Plumier, Descr. Pl. Amér. t. 72. 1693 (according to Howard 1973: 392 and Verdcourt 1996: 11). – Fig. 15 A

Terrestrial herb, delicate, erect, glabrous, annual. Stem 10-30(-40) cm long, green often reddish tinged. Leaves alternate, basely attached; petiole to 0.3-2.5 cm long, glabrous; blade membranous, often drying translucent, broadly elliptic to deltoid, 1-3.7(-5) x 1-3.5(-5.5) cm, apex acute to obtuse, base rounded, cordulate or truncate; palmately 5-7-veined. Inflorescence terminal, solitary; peduncle slender, to 1 cm long, green; spike 1-6 cm long; densely flowered; floral bracts rounded, glabrous, minutely glandular. Fruits somewhat stipitate, globose, brown (young, green ones may be finely longitudinally ridged), stigma apical, a short style may be developed.

Distribution: Widespread in the Neotropics and occurring in the Paleotropics as well; 93 collections studied (GU: 20; SU: 45; FG: 28).

Selected specimens: Guyana: Rupununi Distr., Shea Rock, Jansen-Jacobs *et al.* 4802 (BBS, BRG, U); NW Distr., Polak 155 (NY, U). Suriname: Brokopondo, Sauvain 254 (BBS, P, U); Suriname Distr., Kaipoe weg, UVS 16586 (BBS). French Guiana: Loka, Fleury 148 (CAY, P, US); Cayenne, Macoury, de Granville *et al.* 5522 (CAY, P, U).

Vernacular names: Suriname: kosaka wi (Ndjuka), konsaka wiwiri (Saramaccan); sapoe rakei, eagoe ragoe (Javanese).

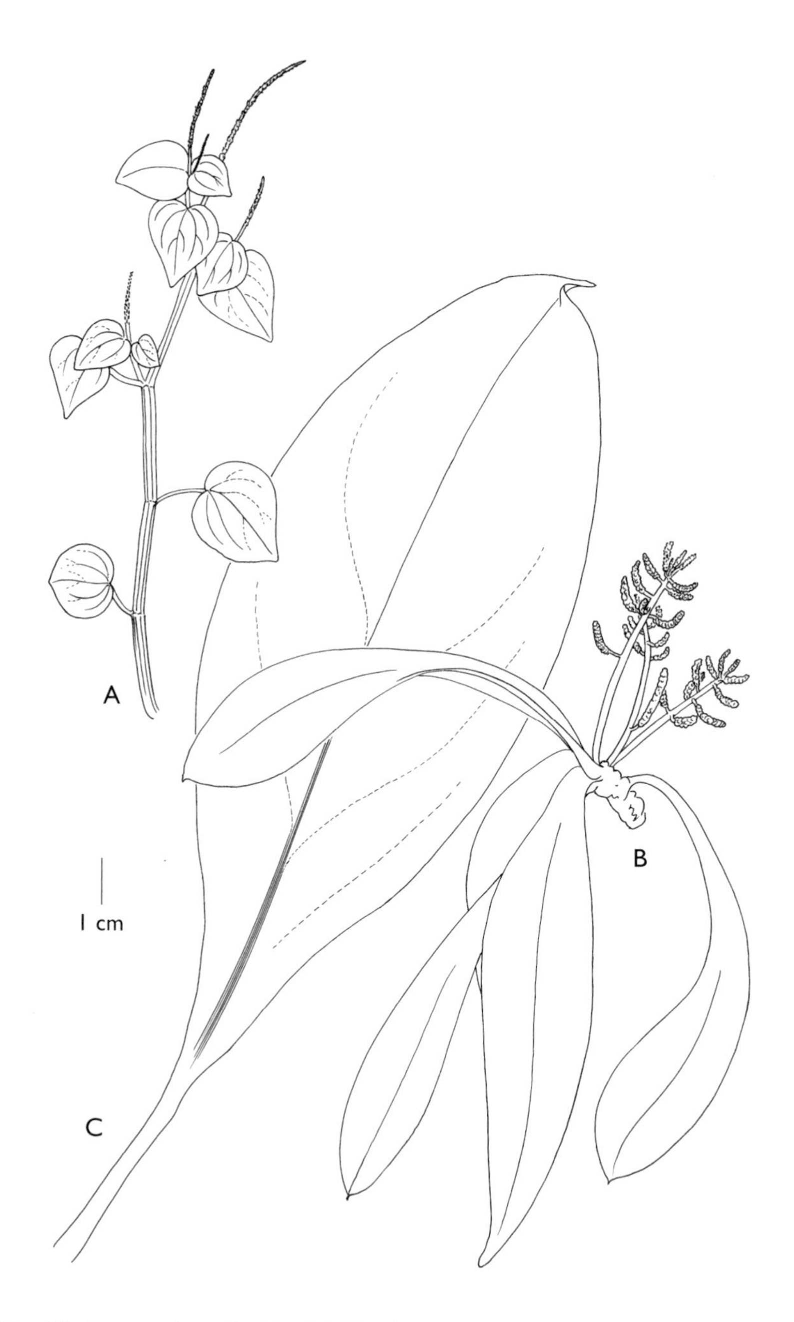

Fig. 15. *Peperomia pellucida* (L.) Kunth: A, habit. *Peperomia pernambucensis* Miq.: B, habit; C, leaf (A, Kramer & Hekking 2858; B, from photograph by T. Schöpke; C, Henkel 4876).

Uses: Suriname: against inflammations, skim bruises, as a diuretic; mixed with other substances against high blood pressure (Raghoenandan, Utrecht internal report). To flavor salads, it can be found in gardens and pots also as a weed. The plants have some medicinal and even poisonous properties (pers. comm. J.C. Lindeman)

Notes: In living specimens the margins of the petiole continue into distinct subnodal ribs, that form a "collar" around the nodes (pers. comm. G. Mathieu).
See note to *Peperomia urocarpa*.

18. **Peperomia pernambucensis** Miq., London J. Bot. 4: 420. 1845. Type: Brazil, Pernambuco, Gardner 1157 (holotype K).
– Fig. 15 B-C

Peperomia paniculata Regel, Bull. Soc. Imp. Naturalistes Moscou 31(2): 542, t. 3. 1859. Type: Brazil, Riedel s.n. (holotype LE, not seen).
Peperomia longifolia C. DC. in A. DC., Prodr. 16(1): 405. 1869. Type: French Guiana, L.C. Richard s.n. (syntype P, not seen); French Guiana, Leprieur s.n. (syntype P, not seen).

Terrestrial or epiphytic, erect herb, glabrous except for pubescent peduncle. Stem to 5 cm long, green. Leaves alternate, basely attached; petiole to 0.25 cm long; blade membranous, narrowly elliptic or oblanceolate, 15-30 x 4-12 cm, apex acuminate, base cuneate; venation pinnate with 4-5(-8) lateral pairs of veins. Inflorescence terminal, paniculate; peduncle 5 cm long, densely velvety pubescent, green; spikes 4-8 cm long, green, densely flowered; floral bracts rounded, glabrous, glandular. Fruits basely attached, cylindrical, slightly verrucose, apex truncate with central stigma.

Distribution: C America, Colombia, Venezuela, Trinidad, Guyana, French Guiana and N Brazil; to 700 m elev.; 30 collections studied (GU: 13; FG: 17).

Selected specimens: Guyana: NW Distr., Waini R., foot Kopinang Falls, de la Cruz 1291 (NY); Kaituma R., Sobai Cr., Fanshawe 2416 (= FD 5152) (NY); Rupununi Distr., Kanuku Mts., Jansen-Jacobs *et al.* 3298 (K, U). French Guiana: Saül, Mori 22724 (CAY, NY, U); Camp Eugène, Bassin du Sinnamary, Cremers *et al.* 13610 (CAY, NY, P, U, US); Station des Nouragues, Cremers *et al.* 10888 (CAY, NY, P, U).

19. **Peperomia popayanensis** Trel. & Yunck., Piperac. N. South Amer. 688, f. 604. 1950. Type: Colombia, Cauca, Lehmann 1003 (holotype GH, isotypes F, NY, none seen). – Fig. 16 A

Terrestrial, ascending, almost glabrous herb, rooting at nodes. Stem glabrous, green. Leaves alternate, slightly subpeltate; petiole to 20 cm long; blade coriaceous, ovate, to 12 x 7 cm, apex acuminate, base rounded to cordulate, minutely ciliate towards apex; somewhat pli-veined with 2 pairs at base and 2-3 pairs from primary vein. Inflorescence axillary, solitary; peduncle slender, 3-7 cm long with 2-3 caducous bracts, green; spike 17 cm long, green; floral bracts, rounded, glabrous. Fruits basely attached, ellipsoid, apex with slender beak and stigma at base of beak.

Distribution: Known from the type collection in Colombia and one collection from French Guiana, that may belong to the species: Sommet Tabulaire, 750 m elev., Cremers 6349 (CAY).

20. **Peperomia purpurinervis** C. DC., Notizbl. Bot. Gart. Berlin-Dahlem 7(62): 496. 1917. Type: Venezuela, Bolivar, Mt. Roraima, Ule 8592 (holotype B, isotype K, photo F). – Fig. 16 B-C

Epiphytic, terrestrial or epilithic herb, creeping to ascending. Stem glabrous to minutely pubescent (with white hairs, Persaud 52), 10-28 cm long, reddish. Leaves alternate, basely attached; petiole 0.2-1 cm long; blade fleshy-coriaceous, elliptic, subobovate or broadly ovate, 1-3.5 x 0.8-2.3 cm, maybe sparsely ciliate, apex acute to obtuse, base acute or obtuse, glabrous or slightly hirtellous, green above, often magenta-red below; veins 3, obvious, often red below. Inflorescence terminal, solitary; peduncle not slender, 0.5-2.5 cm long, glabrous or minutely pubescent; spike erect, 5-7 cm long, yellow or pale green or greenish white, laxly flowered; floral bracts rounded, glabrous, glandular. Fruits basely attached, ovoid to globose, smooth, stigma apical.

Distribution: Venezuela, Guyana and Brazil (Roraima); in dense forest on sandstone, 1120-2750 m elev.; 6 collections studied (GU: 4).

Specimens examined: Guyana: Mt. Roraima, Paikwa Trail, R. Persaud 52 (BRG, K); upper Mazaruni R., Tillett *et al.* 43950 (NY); Merume Mts., Tillett *et al.* 44848 (NY); Potaro-Siparuni region, Mt. Ayanganna, Clarke 9215 (U).

Note: The collection Im Thurn 139 from Mt. Roraima does not represent *Peperomia pennellii* Trel. & Yunck. Although it is a poor specimen it fits in *P. purpurinervis*.

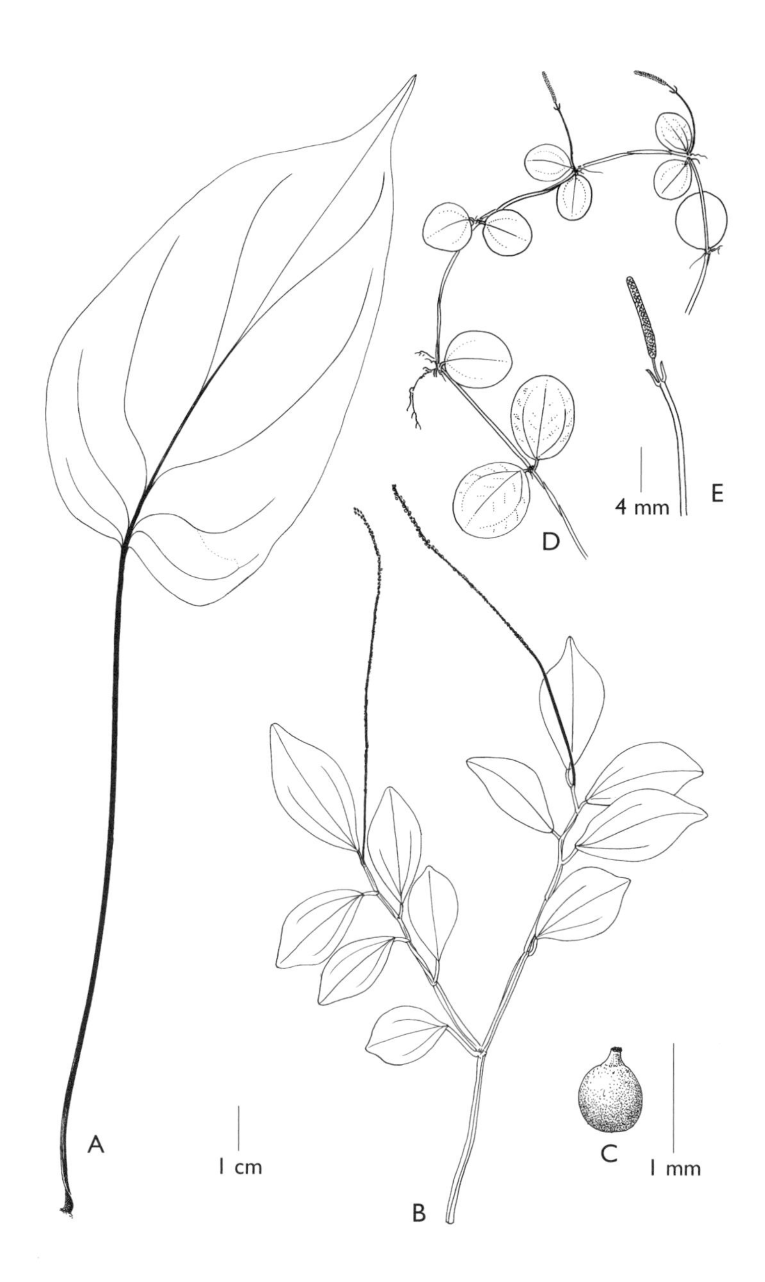

Fig. 16. *Peperomia popayanensis* Trel. & Yunck.: A, leaf. *Peperomia purpurinervis* C. DC.: B, habit; C, fruit. *Peperomia quadrangularis* (J.V. Thomps.) A. Dietr.: D, habit; E, spike on branch tip (A, Cremers *at al.* 6349*;* B-C, Clarke 9215; D-E, Jansen-Jacobs *et al.* 748).

21. **Peperomia quadrangularis** (J.V. Thomps.) A. Dietr., Sp. Pl. 1: 169. 1831. – *Piper quadrangulare* J.V. Thomps., Trans. Linn. Soc. London 9: 202, t. 21, f. 1. 1808. Type: Trinidad, between St. Joseph's and Caroni R. J.V. Thompson s.n. (holotype B-W 738).
– Fig. 16 D-E

Peperomia angulata Kunth in Humb., Bonpl. & Kunth, Nov. Gen. Sp. ed. qu. 1: 66. 1816. – *Piper angulatum* (Kunth) Poir. in Lam., Encycl. Suppl. 4: 468. 1816. Type: Venezuela, Prov. Sucre, Jagua R., Humboldt & Bonpland s.n. (holotype P).
Peperomia muscosa Link, Jahrb. Gewächsk. 1(3): 64. 1820. – *Piper muscosum* (Link)Schult. & Schult. f., Mant. 1: 245. 1822. Type: Brazil, Hoffmannsegg s.n. (holotype B-W 739), syn. nov.

Epiphytic or epilithic herb, creeping or hanging, sometimes mat-forming. Stem quadrangular, minutely pubescent, internodes 2-6 cm long, green brownish or reddish. Leaves opposite, basely attached; petiole 0.1-0.5 cm long, glabrous or minutely pubescent; blade fleshy coriaceous, broadly elliptic, obovate or orbicular, 1-4.5 x 0.8-2.2(-2.7) cm, margin ciliate, apex obtuse, rounded or slightly acute, base acute to rounded, dark green above, with paler veins beneath, minutely puberulent above, sparsely so and glabrescent beneath; palmately 3-veined, veins impressed above, prominent beneath. Inflorescence erect, solitary or two together, axillary or occasionally terminal; peduncle not slender, 0.5-4 cm long, pubescent, 2-bracteate, red, brownish or green; spike up to 6 cm long, green or yellow-green, laxly to moderately flowered; floral bracts rounded or oblong, glabrous, glandular; rachis strongly ridged (at least when dried). Fruits somewhat sunken when young, slightly stipitate when ripe, ovoid, smooth and shiny (*in vivo*) brown, upper half often grayish, verruculose or papillate, with apical stigma.

Distribution: West Indies, Panama and northern S America; in savanna forest, up to 200 m elev.; over 50 collections studied (GU: 34; SU: 15; FG: 4).

Selected specimens: Guyana: Kanuku Mts., Rupununi R., Puwib R., Jansen-Jacobs *et al.* 213 (BRG, NY, U, US); NW Kanuku Mts., Iramaikpan, A.C. Smith 3023 (NY, K, P, U). Suriname: Sipaliwini savanna, Oldenburger *et al.* 497 (BBS, NY, U); Bakhuis Mts., Florschütz & Maas 2783 (BBS, U). French Guiana: Mana R., Baboune Cr., de Granville *et al.* 4695 (CAY, P, U); Grand Inini R., Oldeman B-3519 (CAY, P, U).

Notes: Collections from adjacent Brazil (Pará) had been identified *Peperomia muscosa* Link. There seem to be no significant differences with *P. quadrangularis*. Yuncker (Hoehnea 4: 139. 1974) already suggested that

the two may belong to one species. After having seen the variation in leaf shape and studied the types of *P. muscosa* and *P. quadrangularis*, G. Mathieu and I are convinced that the two are conspecific.
Peperomia quadrangularis could be mistaken for *P. maguirei*. For the differences see under the latter.

22. **Peperomia quadrifolia** (L.) Kunth in Humb., Bonpl. & Kunth, Nov. Gen. Sp. ed. qu. 1: 69. 1816. – *Piper quadrifolium* L., Sp. Pl. ed. 2: 43. 1762. Type: [icon] Plumier, Pl. Amer. ed. Burm. t. 242, f. 3. 1760.

Epiphytic, trailing, glabrous herb. Stem erect to ascending, glabrous, up to 25 cm long, green or reddish. Leaves in whorls of 3-6, usually 4, basely attached; petiole 0.1-0.4 cm long; blade fleshy-coriaceous, subspathulate, or elliptic-, or oblong-obovate, 0.6-1.8 x 0.4-1.0 cm, apex emarginate, occasionally with hairs in indentation, base acute; palmately 3-veined. Inflorescence terminal, solitary, erect; peduncle not slender, up to 1.7 cm long, green; spike 1-3.5 (according to Henkel 946, to 10) cm long, densely flowered, green; floral bracts rounded, glabrous and sparsely glandular. Fruits basely attached, ovoid or subglobose, brown, apex attenuate or somewhat stylose, stigma apical.

Distribution: West Indies, Mexico, C America, Venezuela, Colombia, Guyana, Brazil and Peru; from 1000-2800 m elev.; 2 collections studied (GU: 1).

Specimens examined: Guyana: Pakaraima Mts., Potaro-Siparuni region, Henkel 946 (U).

23. **Peperomia reptans** C. DC., J. Bot. 4: 143. 1866. Type: Colombia, Triana 58 (holotype G-DC, not seen). – Fig. 17 A

Peperomia duidana Trel. in Gleason, Bull. Torrey Bot. Club 58: 354. 1931. Type: Venezuela, Amazonas, slopes of ridge 25, summit of Mt. Duida, Tate 438 (holotype NY, isotype ILL, not seen), syn. according to Mathieu 2001-2006.

Epiphytic, terrestrial or epilithic herb, creeping. Stem succulent, glossy, retrorsely crisp-pubescent, green with red spots or reddish. Leaves alternate, basely attached; petiole 1-1.5(-4.5) cm long, retrorsely appressed-pubescent; blade fleshy, coriaceous, dark green above, light green beneath, broadly ovate to deltoid, 0.8-2.5 x 0.8-2.5 cm, margin ciliate, apex acutely obtuse, base cordate, cordulate or rounded, brown

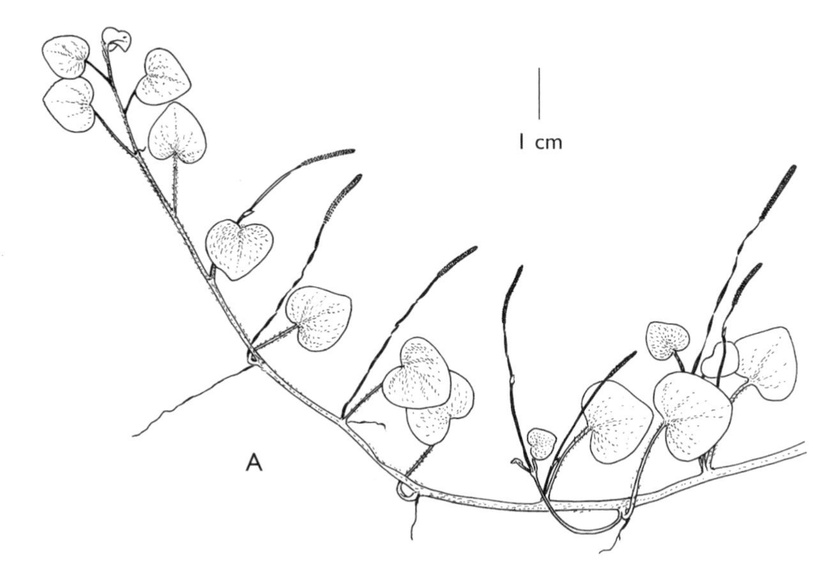

Fig. 17. *Peperomia reptans* C. DC.: A, habit (A, Henkel 4356).

appressed-pilose above, below light green (*Tillett*), red with green *(Persaud)* or brown *(Hahn)* stripes and white hairs *(Persaud)*; venation obvious in dried state, obsoletely palmately 5-7-veined. Inflorescence single, axillary; peduncle not slender, often bracteate, 3-6 cm long, longer than spike, pubescent, sometimes red; spike 1-2.5 cm long, green or yellow, densely flowered; floral bracts rounded, glandular. Fruits basely attached, subglobose to ovoid, with more or less hooked beak and subapical stigma.

Distribution: Colombia, Venezuela and Guyana; from 300-1900 m elev.; 8 collections studied (GU: 8).

Selected specimens: Guyana: Potaro-Siparuni region, Pakaraima Mts., Mt. Wokomong, Henkel 4356, 4434 (U, US); Mt. Roraima, Paikwa Trail, R. Persaud 45 (BRG, K); id., S section, near end of Waruma trail, Warrington *et al.* (=K.E.R.) 43 (K, U); Cuyuni-Mazaruni region, 2 km NW of N tip of "prow" of Roraima, Hahn 5428 (U, US); Mt. Ayanganna, Tillett & Tillett 45132 (NY).

Notes: G. Mathieu has seen the types of *Peperomia reptans* and *P. duidana*, and concluded that these names concern the same species.
P. reptans is very similar to *P. tillettii* Steyerm., but is pubescent while *P. tillettii* is glabrous. The latter is also a species of high altitudes and was described from Venezuela, Edo. Mérida at 1400 m.

24. **Peperomia rhombea** Ruiz & Pav., Fl. Peruv. Chil. 1: 31, t. 46, f. b. 1798. – *Piper rhombeum* (Ruiz & Pav.) Vahl, Enum. 1: 353. 1804. Type locality: Peru, Huanuco; type not designated.

Epiphytic, stoloniferous herb. Stem sparsely pubescent. Leaves in whorls of 3-5, usually 4; petiole 0.1-0.5 cm long, minutely pubescent; blade coriaceous fleshy, subrhombic, elliptic or narrowly elliptic, 1.3-1.5 x 0.4-1 cm, margin ciliolate at apex, obtuse at apex, narrowly acute at base, glabrous above, sparsely minutely pubescent below; distinctly palmately 3-veined. Inflorescence terminal, solitary, simple; peduncle 0.8-3.7 cm long, minutely pubescent; spike 2.2-6 cm long; floral bracts rounded, glabrous, 0.4-0.5 mm in diam. Fruits ellipsoid or ovoid, 0.9-1 mm long, slightly stipitate at maturity, stigma apical.

Distribution: Antilles and northern S America to Peru. No collections reported from the Guianas, but expected to be found.

25. **Peperomia rotundifolia** (L.) Kunth in Humb., Bonpl. & Kunth, Nov. Gen. Sp. ed. qu. 1: 65. 1816. – *Piper rotundifolium* L., Sp. Pl. 30. 1753. pe: [icon] Plumier, Descr. Pl. Amér. t. 69. 1693 (according to Howard 1973: 394 and Verdcourt 1996: 12). – Fig. 18 A-D

Piper nummulariifolium Sw., Prodr. 16. 1788, as 'nummularifolium'. – *Peperomia nummulariifolia* (Sw.) Kunth, Nov. Gen. Sp. ed. qu. 1: 66. 1816, as 'nummularifolia'. Type: Jamaica, Swartz s.n. (holotype S, isotype BM, none seen).
Acrocarpidium nummulariifolium (Sw.) Miq. var. *obcordatum* Miq., Linnaea 20: 118. 1847, as 'nummularifolium' and 'obcordata'. – *Peperomia nummulariifolia* (Sw.) Kunth var. *obcordata* (Miq.) C. DC. in A. DC., Prodr. 16(1): 421. 1869, as 'nummularifolium'. – *Peperomia rotundifolia* (L.) Kunth f. *obcordata* (Miq.) Dahlst., Kongl. Svenska Vetenskapsakad. Handl. 33(2): 101. 1900. Type: Brazil, Martius s.n. (holotype M, not seen).
Peperomia rotundifolia (L.) Kunth f. *ovata* Dahlst., Kongl. Svenska Vetenskapsakad. Handl. 33(2): 101. 1900. – *Peperomia rotundifolia* (L.) Kunth var. *ovata* (Dahlst.) C. DC. in Urb., Symb. Antill. 3: 230. 1902. Type: Brazil, Minas Gerais, Regnell III, 1108 (isotype, M, not seen).
Peperomia bartlettii C. DC., Notizbl. Bot. Gart. Berlin-Dahlem 7(62): 470. 1917. Type: Guyana, Rockstone, Temple bar, Conawarook R., Bartlett 8233 (holotype B, not seen, isotype G, not seen, drawing of type K).
Peperomia lanjouwii Yunck., Acta Bot. Neerl. 6: 393. 1957. Type: Suriname, Nassau Mts., Lanjouw & Lindeman 2598 (holotype U).

Epiphytic or epilithic herb, prostrate with ascending stem tips rooting at nodes. Stem crisp-pubescent when young, internodes 0.3-2 cm long, green often with red spots. Leaves alternate, basely attached, sometimes

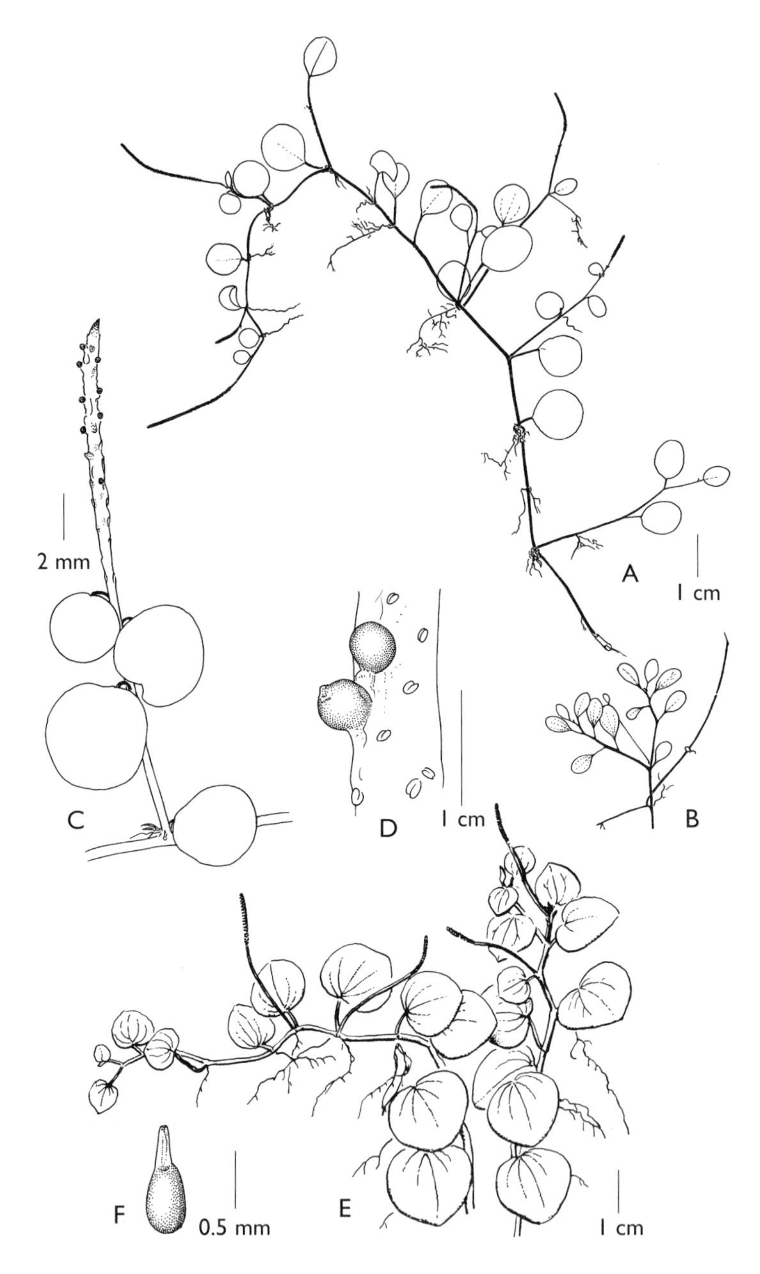

Fig. 18. *Peperomia rotundifolia* (L.) Kunth: A-B, habit; C, detail of branch tip with spike bearing fruits; D, detail spike rachis with fruits. *Peperomia serpens* (Sw.) Loudon: E, habit; F, fruit (A, Henkel 4243; B, Jonker & Daniëls 1136; C, Henkel 5096; D, Clarke 6004; E, Kramer & Hekking 2671; F, Maguire 24092).

subpeltate, fertile branches often with distally progressively smaller leaves; petiole up to 1 cm long, pubescent to glabrous; blades fleshy or fleshy coriaceous, broadly elliptic to elliptic or almost orbicular, (0.5-) 1-1.2 x (0.5-)1-1.2 cm, margin ciliate, apex rounded, base rounded, glabrous to puberulent; obsolete palmately 3-veined. Inflorescence terminal, solitary; peduncle not nodose, without bract, up to 1 cm long, pubescent or glabrous; spike 2-6 cm long, yellow to red, moderately to densely flowered; floral bracts rounded, glabrous, glandular. Fruits (sub)basely attached, globose to ovoid, brown, reticulately verruculose, apex oblique with subapical stigma.

Distribution: Throughout the range of the genus; humid forests from sea level up to 1000 m elev.; ca. 135 collections studied (GU: 51; SU: 43; FG: 41).

Selected specimens: Guyana: Barima-Waini region, McDowell 4368 (K, U); Takutu R., SE Kanuku Mts., Hoffman 456 (U). Suriname: Voltzberg, Jansen-Jacobs *et al.* 6161 (BBS, U); Ulemari, Raghoenandan, UVS 17851 (BBS, U). French Guiana: Saül, de Granville *et al.* 8654 (CAY, NY, P, U); Saül, Feuillet 7388 (CAY, P, US).

Vernacular names: Guyana: follow me (Creole, van Andel 1333); Suriname: ditibi, tisiyi (Saramacca); aneisie wiwiri or pikin-koekoe. French Guiana: wilapili (Wayampi).

Notes: Various authors had recognized formas and varieties on basis of leaf form. With all material available it is clear that *Peperomia rotundifolia* is an extremely variable taxon; specimens may have tiny round leaves, others show round plus somewhat elliptic leaves. As a result, I decided to consider the varieties to be part of one taxon.
Peperomia rotundifolia and *P. serpens*, two very common species, can easily be distinguished by their inflorescences. In the former the inflorescences point out of the mat of vegetation and the spikes are far longer than the peduncle, whereas in *P. serpens* the inflorescences are less conspicuous between the mats and spikes and peduncles are equally long. See also note to *P. delascioi*.

26. **Peperomia serpens** (Sw.) Loudon, Hort. Brit.: 13. 1830. – *Piper serpens* Sw., Prodr. 16. 1788. Type: Jamaica, Swartz s.n. (holotype S, not seen). – Fig. 18 E-F

Epiphytic or epilithic herb, creeping or twining and rooting at nodes. Stem minutely crisp-pubescent, internodes 0.5-3 cm long, green. Leaves alternate, slightly peltate; petiole to 2 cm long, crisp-pubescent; blade fleshy or slightly coriaceous, broadly elliptic or deltoid, 1-2 x 1-2 cm, ciliolate towards apex, apex rounded, base rounded, some hairs on young leaves only; venation obsoletely 3-5-veined. Inflorescence terminal, solitary; peduncle nodose or with 1-2 caducous bracts, not slender, to 3 cm long, crisp-pubescent, not winged, green; spike up to 1.7 cm long, pale green, densely flowered; floral bracts rounded, glabrous, glandular. Fruits basely attached, ellipsoid, brown, apex with slender beak and subapical stigma.

Distribution: Throughout the Neotropics; from sea level to 1200 m; over 160 collections studied (GU: 38; SU: 36; FG: 91).

Selected specimens: Guyana: Shodikar Cr., A.C. Smith 2879 (K, NY); Marudi Mts., Stoffers & Görts-van Rijn *et al.* 220 (BRG, B, NY, U, US). Suriname: Hannover Distr., LBB 15306 (BBS); Brownsberg, Görts-van Rijn & Gouda 504 (BBS, NY, U). French Guiana: junction of Itany R. and Koule-Koule Cr., Feuillet, 2422 (CAY, NY, P, U); Inselberg Telouakem, Tumac Humac Mts., de Granville *et al.* 12134 (B, BBS, BR, CAY, K, NY, P, U, US).

Vernacular names: Suriname: gran koupali (Fleury 174). French Guiana: kaimek pokal (ça sent l'acouti).

Notes: In 1866, a later homonym was published: *Peperomia serpens* C. DC. (J. Bot. 4: 136). It was replaced by *P. dimota* Trel. & Yunck. (1950: 621).
See also note to *P. rotundifolia*.

27. **Peperomia tenella** (Sw.) A. Dietr., Sp. Pl. 1: 153. 1831. – *Piper tenellum* Sw., Prodr. 16. 1788. – *Acrocarpidium tenellum* (Sw.) Miq., Syst. Piperac. 53. 1843. Type: Jamaica, Swartz s.n. (holotype S, isotype K, none seen) – Fig. 19 A-B

Terrestrial, epilithic or epiphytic, not branched, tiny herb, decumbent and rooting at base, fertile stems erect. Stem glabrous to somewhat minutely pubescent, internodes 0.2-1.6 cm long, green with lilac. Leaves alternate, basely attached; petiole 0.1-0.5 cm long, glabrous or with a few hairs at node; blade fleshy-subcoriaceous, elliptic or ovate, 0.6-1.5 x 0.2-1.4 cm, apex obtuse and minutely emarginate with hairs in indentation, base acutish to obtuse, glabrous and ciliate, sometimes sparsely pilose; obsoletely palmately 3-veined. Inflorescence terminal, erect, solitary; peduncle not slender, 0.4-1.5 cm long, glabrous, green; spike 1.7-4 cm long, green, flowers distinctly separate; floral bracts rounded to narrowly oblong, glabrous, glandular. Fruits slightly stipitate at maturity, ellipsoid to obpyriform, with short beak and apical stigma.

Distribution: From Honduras, the West Indies and northern S America; in forests especially cloud forests, at high altitudes up to 1580 m; 8 collections studied (GU: 7).

Selected specimens: Guyana: Mt. Roraima, Im Thurn 140 (BM), 196 (K); Cuyuni-Mazaruni region, Paruima, E edge of Waukauyeng tipu, Clarke 5599 (U); Potaro-Siparuni region, Mt. Ayanganna, Clarke 9136, 9633, 9425 (all U).

Phenology: Flowering throughout the year.

Note: In S Venezuela and Colombia also *Peperomia tenella* var. *tylerii* (Trel.) Steyerm. occurs, which differs in having somewhat larger leaves.

28. **Peperomia tetraphylla** (G. Forst.) Hook. & Arn., Bot. Beechey Voy. 97. 1832. – *Piper tetraphyllum* G. Forst., Fl. Insul. Austr. Prodr. 5. 1786. Type: Society Island, Forster s.n. (holotype GOET, isotype K, none seen). – Fig. 19 C

Piper reflexum L.f., Suppl. Pl. 91. 1781. – *Peperomia reflexa* (L.f.) A. Dietr., Sp. Pl. 1:180. 1831, non *Peperomia reflexa* Kunth 1816. Type: Africa, Cape of Good Hope, Thunberg s.n. (holotype UPS, not seen).

Epiphytic or rarely epilithic herb, cespitose or rooting at nodes and ascending. Stem to 25 cm long, minutely pubescent, green. Leaves in whorls of 4, basely attached; petiole 0.05-0.2 cm long, minutely pubescent; blade fleshy to subcoriaceous, rhombic, ovate-elliptic to suborbicular, 0.6-1.2 x 0.4-1.0 cm, apex obtuse, rounded or emarginate, base obtuse to subacute, somewhat ciliate towards apex and sparsely puberulent to glabrous; obsoletely palmately 3-veined. Inflorescence terminal, solitary; peduncle to 2 cm long, minutely pubescent, green; spike up to 4 cm long, cream-white; rachis pubescent; floral bracts rounded, glabrous, glandular. Fruits basely attached in depressions of rachis, narrowly ovoid to ellipsoid-subcylindrical, apex with short beak and subapical stigma.

Distribution: Pantropical; at higher altitudes; 4 collections studied (GU: 1).

Specimens examined: Guyana, upper slope Mt. Roraima, im Thurn 224 (BM, BRG, K).

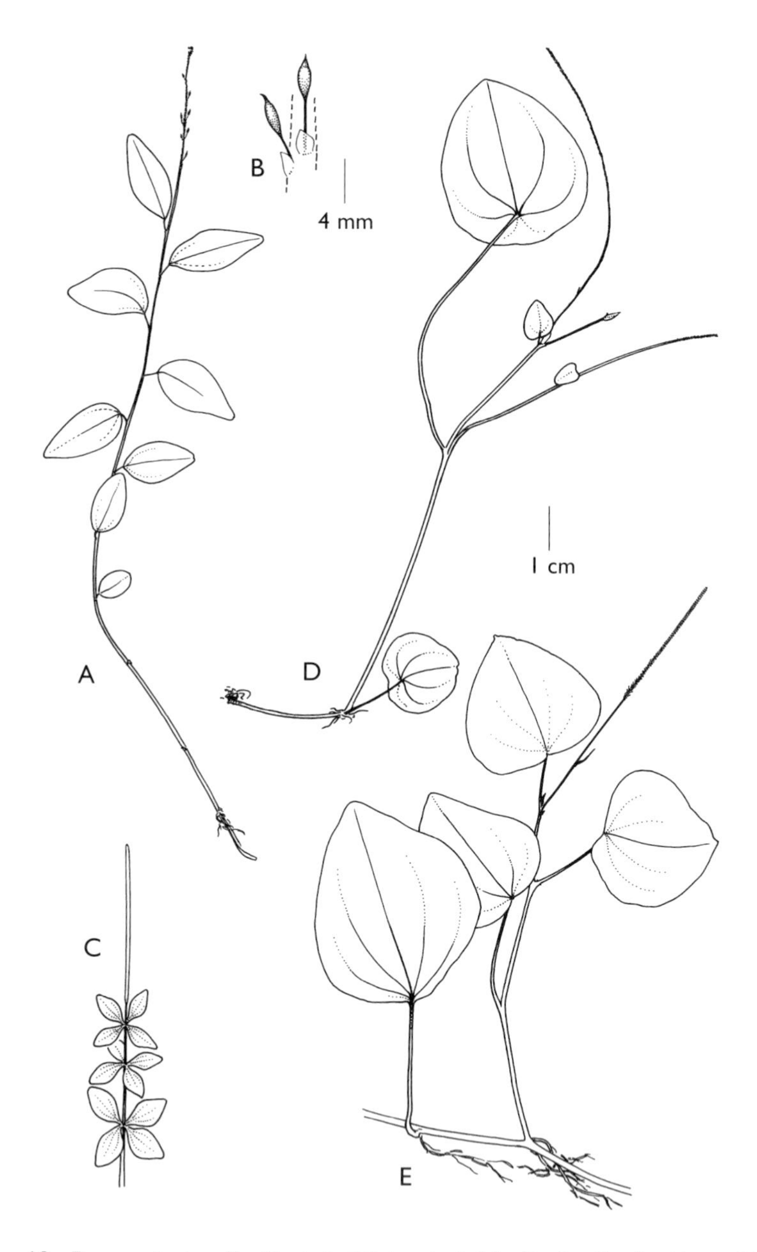

Fig. 19. *Peperomia tenella* (Sw.) A. Dietr.: A, habit; B, detail of spike rachis with stipitate fruits. *Peperomia tetraphylla* (G. Forstr.) Hook. & Arn.: C, habit. *Peperomia transparens* Miq.: D, habit. *Peperomia urocarpa* Fisch. & Mey.: E, habit (A-B, Clarke 5599; C, Im Thurn 224; D, de Granville *et al.* 12132; E, McDowell 2713).

Notes: Dietrich's new combination is illegitimate because of the earlier *Peperomia reflexa* Kunth (Nov. Gen. Sp. ed. qu. 70. 1816) which is a different species.
Peperomia tetraphylla is easily recognizable by the hairy rachis of the spikes and the fruits being sunken in the rachis.
The collection of Im Thurn mentioned above, although sterile, belongs to the species. This Im Thurn collection has been mounted on one sheet with three Brazilian Glaziou collections, two of which also belong to *P. tetraphylla*, the third, however, to a different species.

29. **Peperomia transparens** Miq., Linnaea 20: 121. 1847. Type: Brazil, Minas Gerais, Lo Lobo, Martius s.n. (holotype M, not seen, isotype U). – Fig. 19 D

Epilithic, erect or ascending, glabrous herb, rhizome sometimes tuberous. Stem 10-20 cm long, green or red. Leaves alternate, peltate; petiole attached 0.2-1 cm from base, 1.5-6 cm long; blade fleshy, somewhat bullate *in vivo*, membranous, almost translucent when dried, ovate with acutish apex or broadly ovate to almost orbicular, 2-7(-12) x 1.5-6.5 cm, apex rounded or somewhat acute, base rounded to cordulate; palmately 5-7-veined (veins sometimes purple). Inflorescence terminal, solitary or few together; peduncle slender, 1-4(-15) cm long, red; spike loosely flowered, erect, 2.5-5 cm long, white or green; floral bracts rounded, glabrous, glandular. Fruits basely attached, globose to ovoid, brown, verruculose, apex with slender beak and terminal stigma.

Distribution: Suriname, French Guiana and Brazil (Minas Gerais); on granitic rocks, on 250-850 m elev.; 11 collections studied (SU: 4; FG: 6).

Selected specimens: Suriname: Bakhuis Mts., Florschütz & Maas 2892 (U); Emma Range, Daniëls & Jonker 749 (U). French Guiana: Massif des Emerillons, de Granville *et al.* 3763 (CAY, P, U), Feuillet 1307 (CAY); Mt. Bakra, Cremers 13159 (BBS, CAY, P, U); Inselberg Telouakem, Tumac Humac Mts., de Granville *et al.* 12132 (BBS, CAY, P, U, US).

30. **Peperomia urocarpa** Fisch. & C.A. Mey., Index Sem. Hort. Bot. Petropol. 4: 42. 1838. – *Acrocarpidium urocarpum* (Fisch. & C.A. Mey.) Miq., Syst. Piperac. 60. 1843. Type locality: Brazil; type not designated. – Fig. 19 E

Epiphytic or epilithic herb. Stem rooting at nodes, ascending, crisp-pubescent. glabrescent. Leaves alternate; petiole 1.5-4.5 cm long, crisp-pubescent; blade pale glandular-dotted, deltoid, ovate to broadly ovate, 2-3 x 2.7-7 cm, margin ciliolate towards apex, apex acute to obtuse, base truncate to subcordate, occasionally obtuse, pubescent or villous on both sides, glabrescent; palmately 5-7-veined. Inflorescence solitary, simple, terminal; peduncle 2-4 cm long, crisp-pubescent, 1-2 bracteate; spike 1.5-5 cm long, about as long as peduncle; floral bracts rounded, with scariose margin. Fruits brown, ellipsoid with distinct beak, stigma at base of beak.

Distribution: West Indies, Colombia, Venezuela, Guyana, Suriname, Ecuador and Brazil; from 200-800-1200? m elev.; 13 collections studied (GU: 7; SU: 4; FG: 2).

Specimens examined: Guyana: Cuyuni-Mazaruni region, Paruima, Clarke 1147 (U); along E bank of Ushe R., to Lalluvia Cr. and falls, McDowell 2713 (U). Suriname: 4 km NE of Sipaliwini airstrip, LBB 17284, 17285, 17286 (BBS); Wilhelmina Mts., Juliana peak, Irwin *et al.* 54682 (NY, U). French Guiana: upper Oyapock R., de Granville *et al.* 2522 (P, U); degrad Claude, little Tamouri Cr., de Granville *et al.* 2195 (CAY).

Notes: The collection Pakaraima Mts., NW side Mt. Ayanganna, Cuyuni-Mazaruni region, 1100-1200 m elev., Hoffman 3252 (U) seems to belong to *Peperomia urocarpa* and not - as has been suggested - to *P. manarae* Steyerm. It is a rather glabrous specimen, but bears the typical beaked fruit as in *P. urocarpa. Peperomia manarae* is known only from the Cordillera Costanea in the Aragua province in Venezuela.
Because of the often deltoid leaves, *P. pellucida* and *P. urocarpa* could be confused. *P. urocarpa*, however, has longer leaf blades (2.7-7 cm instead of 1-3.7(-5) cm in *P. pellucida*) and longer petioles (1.5-4.5 cm instead of 0.3-2.5 cm in *P. pellucida*). Fruits in *P. urocarpa* are ellipsoid with a distinct beak, whereas in *P. pellucida* fruits are globose with apical stigma, sometimes a short style has been developed.

2. **PIPER** L., Sp. Pl. 28. 1753.
   Type: P. nigrum L.

   Artanthe Miq., Comm. Phytogr. 32, 40. 1840.
   ≡ Oxodium Raf. 1838
   Enckea Kunth, Linnaea 13: 590. 1840.
   ≡ Gonistum Raf. 1838

Lepianthes Raf., Sylva Tell. 84. 1838.
Type: L. umbellata (L.) Ramamoorthy (Piper umbellatum L.)
Nematanthera Miq., Linnaea 18: 606. 1845 ('1844').
Type: N. guianensis Miq.
Ottonia Spreng., Neue Entdeck. Pflanzenk. 1: 255. 1820.
Type: O. anisum Spreng.
Peltobryon Klotzsch ex Miq., Syst. Piperac. 46. 1843.
Type: not designated
Pothomorphe Miq., Comm. Phytogr. 32, 36. 1840.
≡ Lepianthes Raf. 1838
Quebitea Aubl., Hist. Pl. Guiane 838. 1775.
Type: Q. guianensis Aubl.
Schilleria Kunth, Linnaea 13: 676. 1840 ('1839') (non Schillera Rchb. 1828).
≡ Oxodium Raf. 1838
Steffensia Kunth, Linnaea 13: 609. 1840 ('1839') (non Göpp. 1836).
Type: not designated

Herbs, subshrubs, shrubs or treelets, sometimes with scrambling branches, rarely climbers or lianas. Stems usually with swollen nodes. Prophyll (a bract-like structure) present at base of axillary shoot-apex in flowering branches, minute or well developed, soon caducous; another bract-like sheathing structure is present in some species (e.g. *P. hispidum*). Leaves alternate; petioles often vaginate or winged, at base only or to apex (base of leaf blade), their margins may extend beyond base of blade; blades basely attached or peltate, symmetrical or asymmetrical, at base often quite some difference in attachment of blade to petiole (measures are given where this occurs), glabrous or variously pubescent, often glandular-dotted, sometimes scabrous; palmately or pinnately veined. Inflorescences simple, pedunculate spikes, solitary, leaf-opposed (in *P. peltatum* axillary on reduced, leafless branches, umbel-like); floral bracts peltate, rounded, triangular or cucullate, subtending reduced flowers. Flowers bisexual (in *P. hymenophyllum* protandrous, functionally unisexual); stamens 1-6, filaments short (long exserted in *P. hymenophyllum*); ovary sessile or occasionally on a short stipe, stigmas (2-)3-4, sessile or on a style. Fruits drupes, embedded in rachis to exserted, separate from each other or in compact rows, glabrous or with indument, globose or variously shaped due to compression, stigmas persistent.

Distribution: Pantropical, ca. 2000 species; in the Neotropics ca. 600 species; in the Guianas 58 species of which 11 endemics; weedy species in secondary vegetation, most species are found in the understory of various types of forest, in open spaces or on dark, humid places; from sea level to 2000 m.

Note: Several authors have arranged the many species into separate genera, or in various subgeneric entities. Due to the fact that in this

treatment only a relatively small area is covered, I have not gone into the trouble of assigning species to subgeneric groups. There is no general agreement on the characteristics that describe those groups. It is better to await the results of recent cladistic and phylogenetic research that will elucidate the relationships within Piperaceae and especially in *Piper*.
One of the characters used to define sections is the number of stamens per flower. I am convinced that this is constant, it is given in literature (Jaramillo & Manos) that this number depends on the tightness of the floral arrangement on the spike. Only if we find the same combination of character states in different parts of the plant it could be useful in a flora, especially in the key. In this treatment I have tried to use practical characters instead.
In dicotydonous plants two prophylls can be developed, but in *Piper* only one is developed. The place of *Piper* in either dicots or monocots is under study.

KEY TO THE SPECIES AND INFRASPECIFIC TAXA

1 Leaf blade distinctly peltate; inflorescence axillary (but in fact on a terminal, reduced leafless shoot), umbel-like . . . . . *45. Piper peltatum*
Leaf blades basely attached; inflorescences leaf-opposed, simple 2

2 Leaves palmately veined or palmately-pinnately veined with secondary veins originating from the lower 1/4 of the primary vein 3
Leaves pinnately veined, i.e. secondary veins originating from the lower 1/3 to throughout the primary vein 7

3 Veins originating from the base and 1 or 2 branching off the lower 1/4 of the primary vein 4
Veins all originating from the base 5

4 Leaf base acute; cultivated . . . . *42. P. nigrum*
Leaf base rounded to cordate; in natural habitats . . . . *49. P. poiteanum*

5 Primary veins 9-11, leaf base deeply cordate, margin ciliate . . . . *40. P. marginatum*
Primary veins 5-7(-9), leaf base acute or rounded to subcordate, margin not ciliate 6

6 Stem without spines; tertiary venation obscure . . . . *5. P. amalago*
Stem often with spines opposite petiole; tertiary venation prominent below . . . . *52. P. reticulatum*

7(2) Leaf blades less than 3 cm wide 8
Leaf blades more than 3 cm wide 13

8 Leaf base rounded to subcordate . . . . *18. P. consanguineum*
Leaf base obtuse to acute 9

9 Secondary veins 2-5 per side, originating from the lower 1/2 of the primary vein; floral bracts densely marginally fringed 10
Secondary veins 6-15 per side, originating throughout the primary vein; floral bracts glabrous 11

10 Stem glabrous; leaf blade not scabrous above; secondary veins 2-3 per side; reported from high elevation . . . . *13. P. bolivaranum*
Stem villous; secondary veins ca. 5 per side; leaf blade somewhat scabrous above; known only from type collection, not from high elevation . . . . *55. P. salicifolium*

11 Stem retrosely pubescent; secondary veins 6-10 per side . . . . *6. P. angustifolium*
Stem glabrous except for subnodal lines of hairs; secondary veins 8 or more per side 12

12 Leaf blade 3-4 x as long as wide, apex acute to shortly acuminate; secondary veins 8-12 per side . . . *7a. P. anonifolium* var. *anonifolium*
Leaf blade 4-6 x as long as wide, apex long-acuminate; secondary veins 8-15 per side . . . . *27. P. eucalyptifolium*

13(7) Leaf blades 14-35 cm wide, deeply lobed at base 14
Leaf blades less than 14 cm wide, or if wider then not distinctly lobed 15

14 Fruits puberulent at apex . . . . *16. P. cernuum*
Fruits subglabrous . . . . *43. P. obliquum*

15 Stems and leaves glabrous, but leaf margins ciliate 16
Another combination of characters for stems, leaves and margins 17

16 Secondary veins 10-16, originating throughout the primary vein . . . . *9. P. augustum*
Secondary veins 4-6, originating from the lower 1/2 of the primary vein . . . . 17. *P. ciliomarginatum*

17 Whole plant densely dark glandular dotted; stem glabrous; leaves sparsely villous on both sides . . . . *26. P. dumosum*
Plants not dark glandular dotted; stems with subnodal lines of hairs or glabrous or variously pubescent; leaves glabrous or variously pubescent 18

18 Stems with subnodal lines of hairs 19
Stems glabrous or variously pubescent 20

19 Leaf blade glabrous below . . . . *7a. P. anonifolium* var. *anonifolium*
Leaf blade pubescent below . . . . *7b. P. anonifolium* var. *parkerianum*

20 Leaf base very unequally attached to petiole, difference 0.5-1(-3) cm 21
Leaf base equal or slightly unequally attached to petiole, to less than 0.5 cm 22

21 Petiole without stalked glands (tubercles); leaf apex acute or acuminate . . . . *8. P. arboreum*
Petiole with stalked glands (tubercles); leaf apex obtuse, rounded or slightly acute . . . . *57. P. tuberculatum*

22 Leaf blades - at least the veins - variously pubescent 23
Leaf blades glabrescent or glabrous on both sides 54

23 Leaf blades pubescent on both sides - at least on the veins - and/or blades scabrous 24
Leaf blades glabrous above, pubescent below - at least on the veins - blades never scabrous 35

24 Leaf blades scabrous, at least above 25
Leaf blades not or hardly scabrous 30

25 Spike distinctly curved . . . . *2. P. aduncum*
Spikes straight, not curved 26

26 Secondary veins originating from the lower 1/3 of the primary vein; veins strongly prominent below; savanna shrub . . . . *30. P. fuligineum*
Secondary veins originating throughout or from the lower 1/2 of the primary vein; veins at most prominent below; in forest or secondary vegetation 27

27 Petiole and peduncle 1.5-2 cm long; leaf blade broadly ovate . . . . *51. P. remotinervium*
Petioles and peduncles less than 1.5 cm long; leaf blades lanceolate-oblong, elliptic to ovate or rhombic 28

28 Herb or subshrub, 0.15-1.5 m tall; secondary veins originating throughout the primary vein . . . . *15. P. brownsbergense*
Shrubs or treelets, 2-5 m tall, occasionally with scrambling branches; secondary veins originating from the lower 1/2 of the primary vein . . 29

29 Prophyll densely pubescent; leaf blade rhombic to subobovate; spike 7-8 cm long; anthers dehiscing laterally; fruits glabrous with some hairs at top . . . . *23. P. dilatatum*
Prophyll glabrous; leaf blade elliptic or elliptic-ovate; spike 8-14 cm long; anthers dehiscing transversely apically; fruits hirsute . . . . *34. P. hispidum*

30(24) Spikes erect 31
Spikes pendent 33

31 Peduncle (filiform) much longer than the spike; spike 1-1.7 cm long . . . . *32. P. guianense*
Peduncle (not filiform) shorter than the spike; spike 7-20 cm long . . . 32

32 Petioles 0.5-1.5 cm long; spikes 7-8 cm long; leaf blade elliptic, rhombic or subobovate . . . . *23. P. dilatatum*
Petioles 1.5-2 cm long; spikes 10-20 cm long; leaf blade broadly ovate . . . . *51. P. remotinervium*

33 Leaf blade margin hardly ciliate, base almost equally attached to petiole; floral bracts cucullate, pilose at inner side . . . . *21. P. cyrtopodum*
Leaf blade margin ciliate, base unequally attached to petiole difference 0-0.8 cm; floral bracts not cucullate, marginally fringed 34

34. Stem pilose with 2 mm long hairs; internodes drying finely ridged . . . . *15. P. brownsbergense*
Stem hirsute with 2.5 mm long hairs; internodes not ridged . . . . *33. P. hirtilimbum*

35(23) Leaf apex obtuse or acutish . . . . *36. P. humistratum*
Leaf apices acute to long-acuminate 36

36 Leaf blade slightly bullate . . . . *44. P. paramaribense*
Leaf blades smooth 37

37 Peduncle (filiform) much longer than the spike; leaf base equally attached to the petiole; secondary veins 4-5(-7) per side, at base originating with 2 pairs . . . . *32. P. guianense*
Peduncle (stout) as long as or far shorter than the spike; without above given combination of characters 38

38 Scandant or creeping shrub; floral bracts densely marginally fringed; spike conspicuously protandrous, conical in male stage, stamens long exserted . . . . *37. P. hymenophyllum*
Herbs, shrubs or treelets; floral bracts glabrous or variously pubescent or fringed; spikes in male stage cylindrical, stamens short 39

39 Floral bracts cucullate, glabrous or sparsely pilose 40
Floral bracts trigonous or rounded-trigonous, densely marginally fringed 44

40 Low herb or subshrub, sometimes creeping; secondary veins conspicuously anastomosing . . . . *18. P. consanguineum*
Shrubs, never creeping; secondary veins not conspicuously anastomosing 41

41 Secondary veins 5-7 per side, originating from the lower $^{2}/_{3}$ of the primary vein . . . . *31. P. glabrescens*
Secondary veins 7-12 per side, originating throughout the primary vein 42

42 Leaf base subcordate, obtusish, rounded, or cuneate, not lobed . . . . *44. P. paramaribense*
Leaf base unequally cordate, with one lobe more or less covering the petiole 43

43 Spike pendent, 1-1.5 cm, in fruit 2.5 cm long; ovary and fruit stylose . . . . *14. P. brasiliense*
Spike erect, 3.5-4.5 cm long; ovary and fruit not stylose . . . . *22. P. demeraranum*

44(39) Spikes pendent 45
Spikes erect 47

45 Petiole 2.5-4.5(-7.5) cm long; leaf base strongly asymmetrical; secondary veins widely spaced . . . . *38. P. inaequale*
Petiole 0.2-1.3 cm long; leaf base unequal; secondary veins not widely spaced 46

46 Indument of spreading, to 0.5 mm long, hairs; leaf apex long-acuminate; fruits distinctly separate at maturity, 2.5-4 mm in diam. . . . . *1. P. adenandrum*
Indument of short crisp, to 0.2 mm long, hairs; leaf apex short-acuminate or acute; fruits not distinctly separate, to 2 mm in diam. . . . . *11. P. avellanum*

47 Petiole conspicuously winged; leaf blade hirsute below . . . *25. P. duckei*
Petioles at most vaginate; leaf blades variously pubescent below 48

48 Leaf blades crisp- to erect-pubescent below 49
Leaf blades appressed pubescent below 51

49 Leaf base acute to cuneate . . . . *58. P. wachenheimii*
Leaf base obtuse, rounded or subcordate 50

50 Leaf apex acute; fruits glabrous . . . . *29. P. flexuosum*
Leaf apex long-acuminate; fruits hirsute . . . . *56. P. trichoneuron*

51 Secondary veins originating at an angle of 60° and abruptly curving upwards at an angle of 120°, well within the margin . . . . *35. P. hostmannianum*
Secondary venation not as above 52

52 Petiole 0.3-1 cm long; leaf blade somewhat scabrous and somewhat bullate; peduncle crisp-pubescent .... *50. P. pulleanum*
Petioles 0.5-6.5 cm long; leaf blade not scabrous, nor bullate; peduncle glabrous 53

53 Stipules about as long as petiole; leaf blade broadly ovate, not conspicuously glandular dotted below .... *19. P. coruscans*
Stipules longer than petiole; leaf blade elliptic-ovate, densely glandular-dotted below .... *46. P. perstipulare*

54(22) Fruits with 4 apical projections or fruits distinctly winged, ridged or sulcate 55
Fruits without 4 apical projections, not distinctly winged nor ridged nor sulcate 57

55 Fruits with 4 prominent apical projections .... *4. P. alatabaccum*
Fruits distinctly winged, ridged or sulcate 56

56 Infructescence 2.5-4.5 cm long, fruits distinctly sulcate .... *10. P. aulacospermun*
Infructescence 10-12 cm long, fruits ridged or winged .... *12. P. bartlingianum*

57 Ovaries and fruits distinctly stylose 58
Ovaries and fruits with sessile stigmas, occasionally with inconspicuous short style 59

58 Leaf blade not conspicuously glandular; spike 12-15 cm long .... *20. P. crassinervium*
Leaf blade densely conspicuously glandular; spike 2-4 cm long .... *47. P. phytolaccifolium*

59 Climbing or creeping herb; leaf base cordate or obtuse; flowers protandrous or unisexual, stamens long exserted .... *37. P. hymenophyllum*
Subshrubs, shrubs or treelets; leaf base acute, obtuse or rounded; flowers bisexual, stamens short, not exserted 60

60 Leaf base unequally attached to the petiole difference 0.5-1 cm .... *8. P. arboreum*
Leaf bases less unequal, difference at the petiole less than 0.5 cm 61

61 Floral bracts not marginally fringed, glabrous or at most somewhat pilose 62
Floral bracts densely marginally fringed 64

62 Spike pendent, 1 cm long .... *53. P. rudgeanum*
Spikes erect, to 7 cm long 63

63 Petiole up to 3.2 cm long; leaf blade more than 7 cm wide; secondary veins from lower 2/3 of primary vein . . . . *31. P. glabrescens*
Petiole up to 2 cm long; leaf blade less than 6.5 cm wide; secondary veins originating from throughout primary vein . . . *48. P. piscatorum*

64 Leaf blades asymmetrical, i.e. one half wider and more obtuse at least below the middle 65
Leaf blades symmetrical 67

65 Spike erect, to 10 cm long; veins often drying yellowish . . . *3. P. aequale*
Spikes pendent, to 6 cm long; veins not drying yellowish 66

66 Petiole 1-2.2 cm long; leaf blade dark glandular-dotted; peduncle 0.5-1.5 cm long . . . . *24. P. divaricatum*
Petiole 2.5-4.5(-7.5) cm long; leaf blade not conspicuously glandular-dotted; peduncle 1.5-2.5 cm long . . . . *38. P. inaequale*

67 Leaf base acute, cuneate or obtusish 68
Leaf base truncate, obtuse or rounded 71

68 Secondary veins originating from lower 1/2 to lower 2/3 of primary vein; peduncle up to 3 cm long, spikes 5-9 cm long 69
Secondary veins originating from lower 3/4 or throughout primary vein; peduncle up to 2 cm long, spikes 1-5 cm long 70

69 Leaf blade conspicuously glandular-dotted; secondary veins 4-6 per side . . . . *39. P. insipiens*
Leaf blade not conspicuously glandular-dotted; secondary veins 3-4 per side . . . . *54. P. rupununianum*

70 Petiole 1-2 cm long; spike erect; fruits puberulent . . . . *28. P. fanshawei*
Petiole 0.4-0.8 cm long; spike pendent; fruits glabrous . . . . *41. P. nigrispicum*

71 Spike 1-1.7 cm long . . . . *32. P. guianense*
Spikes 7-11 cm long 72

72 Stipules about as long as petiole; leaf blade broadly ovate, base truncate to cordulate, not conspicuously glandular-dotted below . . . . *19. P. coruscans*
Stipules longer than petiole; leaf blade elliptic-ovate, base obtuse to rounded, densely glandular-dotted below . . . . *46. P. perstipulare*

1. **Piper adenandrum** (Miq.) C. DC. in A. DC., Prodr. 16 (1): 273. 1869. – *Artanthe adenandra* Miq., Syst. Piperac. 515. 1844. Type: French Guiana, Leprieur 152 (holotype G-DEL, isotype G-DC).
– Fig. 20 A-C

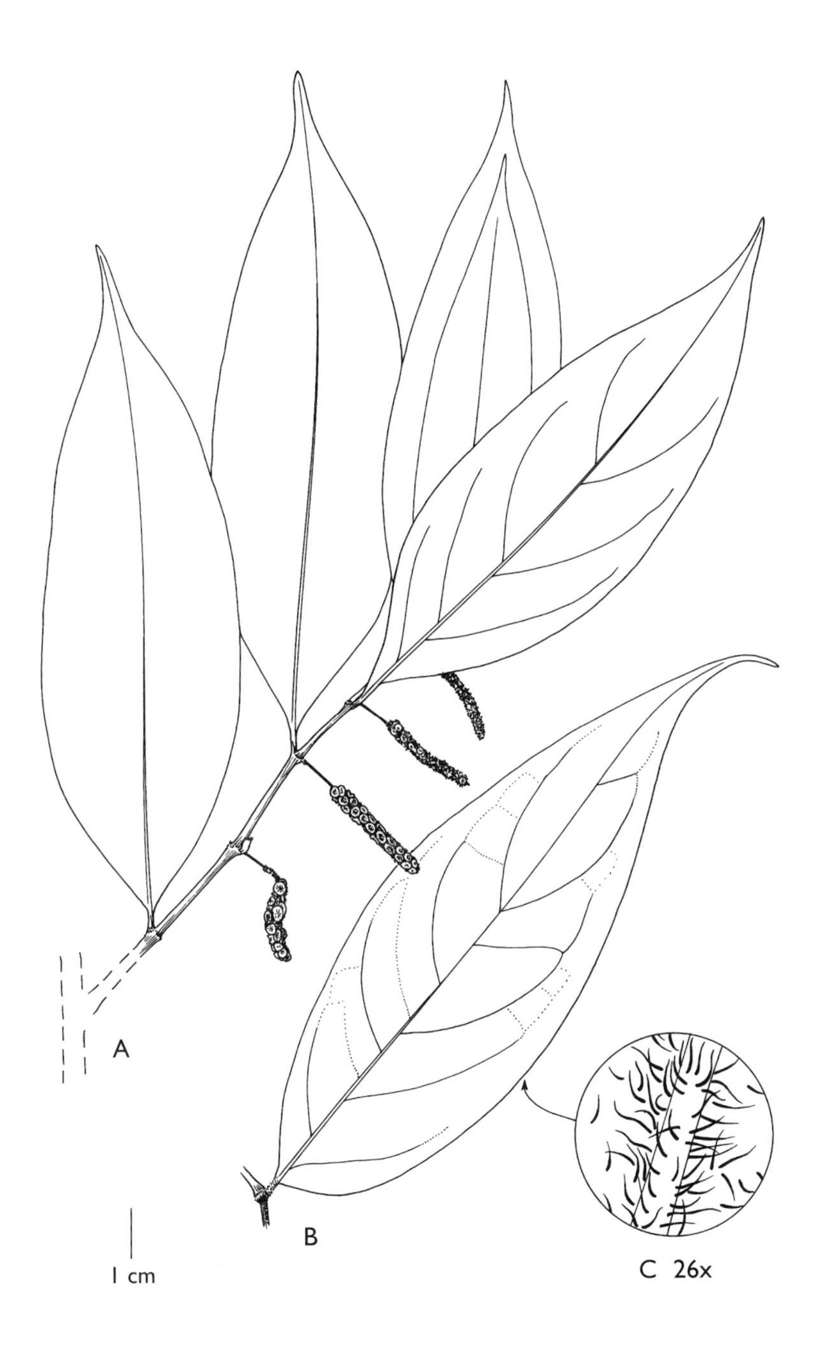

Fig. 20. *Piper adenandrum* (Miq.) C. DC.: A, habit; B, leaf; C, detail of indument on lower leaf surface (A, Hoff 6799; B-C, Wallnöfer 13475).

Piper oblongifolium sensu C. DC. (1869: 273) as to description and specimen cited, non (Klotzsch) C. DC.

Shrub or treelet, 1.5 m tall, pubescent or villous, hairs to 0.5 mm long; upper internodes densely glandular. Petiole 0.2-0.5(-1) cm long, vaginate to middle; blade not conspicuously glandular-dotted, lance-elliptic or elliptic-ovate, 8-16.5 x 3-6(-7) cm, margin ciliate, apex long-acuminate, base unequally attached to petiole difference 0-0.2 cm, acute or obtusish, glabrous or pubescent above, (densely) crisp-pubescent or villous beneath; pinnately veined, secondary veins 5-6 per side, originating from lower 3/4 of primary vein. Inflorescence pendent; peduncle 0.5-1.1 cm long, sometimes reddish tinged, crisp-pubescent or villous; spike 1-5 cm long, green to pale or dark yellow, apiculate; rachis glabrous; floral bracts densely marginally fringed; anthers laterally dehiscent. Fruits separate at maturity, depressed globose, 2.5-3(-4) mm in diam., verruculose; stigmas 3, sessile or fruits somewhat stylose.

Distribution: Venezuela and the Guianas; from sea level to 500 m elev.; 81 collections studied (GU:14; SU: 5; FG: 62).

Selected specimens: Guyana: Upper Takutu - Upper Essequibo region, Gunn's, Clarke 7059 (U, US); Acarai Mts., Clarke 2947 (U, US). Suriname: Gran Santi, Sauvain 448 (BBS, U); Lely Mts., Lindeman & Stoffers *et al.* 834 (BBS, NY, U). French Guiana: Saül area, foot Mt. Galbao, de Granville *et al.* 8398 (CAY, K, NY, P, U, US); Kaw Mts., Feuillet 2240 (CAY, P, U, US).

Vernacular names: Guyana: tona tonakeng (Carib; van Andel, pers. comm.). Suriname: katun uwi (Sauvain 448). French Guiana: ampukuwiwii (Boni; Fleury 833).

Notes: *Piper adenandrum* and *P. avellanum* had been considered conspecific by Kramer & Görts-van Rijn (1968: 418). There are, however, several differentiating characters: *P. adenandrum* has long-acuminate blades, indument of spreading, to 0.5 mm long hairs and distinctly separate fruits, 2.5-3(-4) mm in diam., whereas *P. avellanum* blades are (short) acuminate or acute, the hairs are crisp, 0.2 mm long, and the fruits 2 mm in diam., not distinctly separate.

2. **Piper aduncum** L., Sp. Pl. 29. 1753. – *Steffensia adunca* (L.) Kunth, Linnaea 13: 633. 1840 ('1839'). – *Artanthe adunca* (L.) Miq., Syst. Piperac. 449. 1844. Type locality: Jamaica; type not designated.
– Fig. 21 A

Fig. 21. *Piper aduncum* L.: A, habit (A, Pulle 9).

Shrub or treelet, 3-7 m tall. Stem sparsely pubescent to glabrescent; upper internodes rather slender. Prophyll to 2.5 cm long, pubescent at least on primary vein. Leaves somewhat glandular-dotted; petiole 0.3-0.8 cm long, pubescent, vaginate near base; blade scabrous, yellow-green, lanceolate or lanceolate-oblong or elliptic-ovate, 12-24 x 3-8 cm, apex acuminate, base unequally attached to petiole 0.3-0.4 cm, rounded to subcordate, sparsely pubescent at least on veins above or glabrescent, sparsely pubescent or glabrescent below, somewhat rugulose; pinnately veined, secondary veins 6-8 per side, arising from $^{1}/_{2}$ to $^{3}/_{4}$ of primary vein, sharply ascending, impressed above, prominent below, tertiary

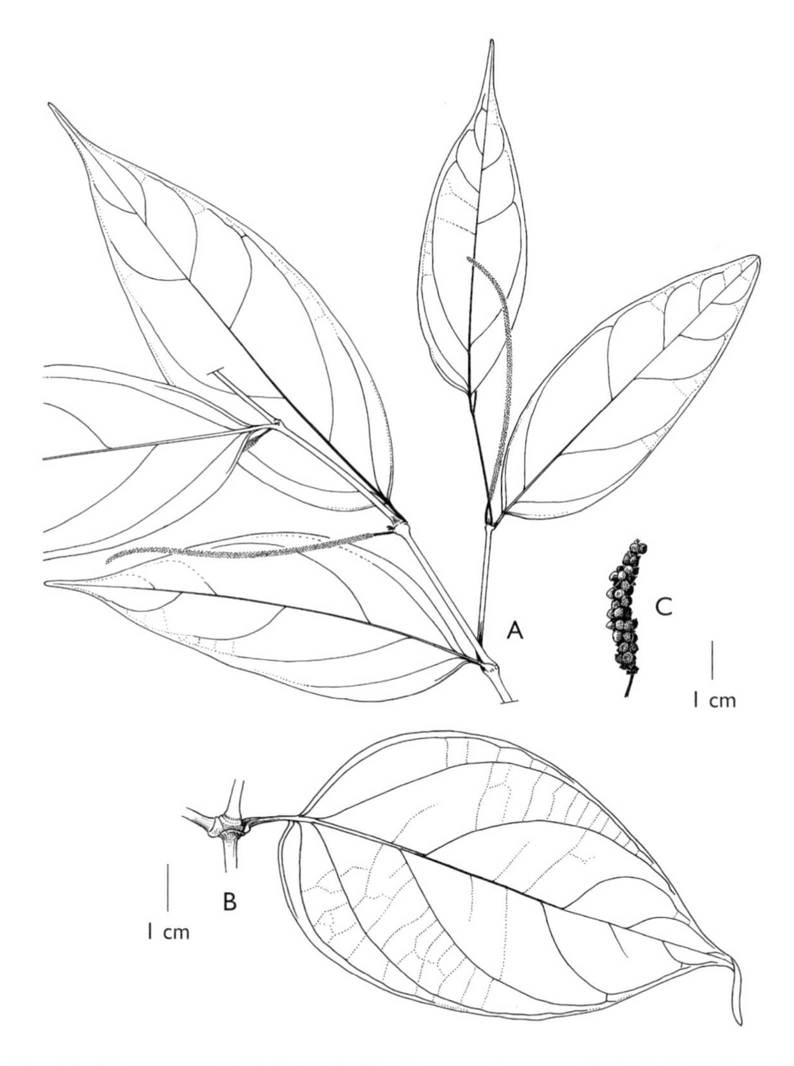

Fig. 22. *Piper aequale* Vahl: A, habit; B, part of stem with leaf; C, spike with fruits (A-B, Lent 3360; C, Feuillet 2474).

veins parallel between secondaries, minor veins reticulate. Inflorescence recurved; peduncle 0.7-1.2(-2.5) cm long, sparsely pubescent; spike 8-15 cm long, white to green, apiculate; flowers densely arranged in rings or spirals; floral bracts densely marginally fringed; stamens with broad connective. Infructescence recurved; fruits obovoid, trigonous or somewhat tetragonous, 0.7-0.8 mm in diam., glabrous or glabrescent, glandular, stigmas 3, sessile.

Distribution: Throughout the range of the family in the Neotropics; in open or somewhat shady places; over 100 collections studied (GU: 33; SU: 38; FG: 35).

Selected specimens: Guyana: Mazaruni Station, Jenman 5290 (BM, BRG, NY); CAS Mon repos, Harrison 1746 (BRG, K, NY). Suriname: s.l., Maguire & Stahel 27171 (NY, P); Wanica, Kramer & Hekking 2605 (BBS, NY). French Guiana: Piste de St. Elie, Hahn 3784 (CAY, NY, P, U); Cabassou Cr., Hoff 5399 (CAY, P, U).

Vernacular names: Guyana: bushi man lembelembe (Hugh-Jones 132). French Guiana: ga(an) man udu (ana) (Ndjuka, several collections by Sauvain and Fleury); watnle (Wayana; Fleury 1415).

Note: *Piper aduncum* is easily recognizable by the strongly curved spikes, and scabrous leaves that are often yellow-green and drooping.

3. **Piper aequale** Vahl, Eclog. Amer. 1: 4, t. 3. 1797. Type: Montserrat, Ryan, s.n. (holotype C, isotypes BM, U). – Fig. 22 A-C

Piper jericoense Trel. & Yunck., Piperac. N. South Amer. 397, f. 359. 1950. Type: Colombia, Alejandro 2730 (holotype US, isotypes F, ILL, none seen).

Shrub or treelet, 1-4 m tall, glabrous. Prophyll 10-15 mm long, slender, acute. Petiole slender, 1-1.5(-4) cm long, vaginate only at base; blade membranous or subcoriaceous, not conspicuously glandular-dotted, shiny green above, pale below, often drying yellowish green, lanceolate to lance-elliptic or elliptic-ovate, broadly ovate on sterile branches, often asymmetrical: with one side wider in the lower half, 12-23 x 3.5-9(-11) cm, apex acuminate, base equally attached to petiole, acute, cuneate or obtuse, may be truncate or subcordate on sterile branches; pinnately veined, secondary veins 4-6 per side, originating at a steep angle from 3/4 or more of primary vein, 2-3 lower ones from near base, slightly prominulous on both sides, often pale or yellow, tertiary venation inconspicuously reticulate, with some more prominent ones transverse. Inflorescence erect; peduncle slender, 1-1.5 cm long; spike to 10 cm long, white or green, apiculate or not; rachis glabrous; flowers densely crowded; floral bracts marginally fringed. Infructescence (grey-)green; fruits oblongoid to trigonous, 0.7-1(-3) mm in diam., brownish green, glabrous, stigmas 3, sessile.

Distribution: West Indies, C and S America; in forest or even in secondary vegetation, often on granitic outcrops; from sea level to 1500 m elev.; over 100 collections studied (GU: 26; SU: 30; FG: 49).

Selected specimens: Guyana: Barima-Mazaruni region, Pipoly 8291 (BBS, NY, U); Mt. Iramaikpang, NW slope Kanuku Mts., A.C. Smith 3677 (NY, P, U). Suriname: Tumac Humac Range, Inselberg Telouakem, de Granville *et al.* 12121 (BBS, CAY, U); Wilhelmina Mts., Juliana Top, Irwin *et al.* 54610 (NY, U). French Guiana: Saül area, Görts-van Rijn 54 (CAY, NY, U, US), foot Mt. Galbao, de Granville *et al.* 8399 (CAY, P, U, US).

Vernacular names: French Guiana: tumowato (Carib), tona tona keng (Carib; this name is not unique for this taxon); yalitakuã sili, yakamileni piã (Wayampi).

Notes: This glabrous species is characterized by the equal leaf base and slender petiole, but usually distinctly asymmetrical lower half of blade; the leaves dry yellowish green.
Unlike Steyermark (1984: 312), I do not accept varieties, knowing how variable the shape of the leaves can be even on the same plant.

4. **Piper alatabaccum** Trel. & Yunck., Piperac. N. South Amer. 408, f. 371. 1950. Type: Guyana, Essequibo R., Tiger Cr., Labbakabra Cr., Sandwith 1191 (holotype NY, isotypes K, U). – Fig. 23 A-C

Shrub or treelet, 2-3 m tall, glabrous. Petiole 1.2-2.5 cm long, vaginate at base; blade not conspicuously glandular-dotted, elliptic, 20-31 x 6-12 cm, apex long-acuminate, base equally attached to petiole, acute; pinnately veined, secondairy veins 7-10 per side, originating from throughout primary vein. Inflorescence erect; peduncle to 1.2 cm long; spike 5-8 cm long, to 12 cm in fruit, white turning green, not apiculate; floral bracts small, rounded peltate, glabrous, caducous; rachis brownish crisp-pubescent. Fruits separate, obovoid with 4 prominent, deltoid or lobed projections at top, glabrous or papillose, stigmas 4, sessile.

Distribution: Venezuela (Bolivar), the Guianas; from understory of mixed forest (on slopes); to 750 m elev.; 40 collections studied (GU: 10; SU: 8; FG: 22).

Selected specimens: Guyana: Mt. Makarapan, Maas *et al.* 7490; Essequibo region, W Demerara, Henkel 418 (U, US). Suriname: Brownsberg Nature Reserve, Webster 24089 (BBS, U); Tumac Humac Mts., Talouakem Inselberg, de Granville *et al.* 12324 (CAY, P, U, US). French Guiana: Upper Inini, de Granville *et al.* 3978 (CAY, U); Trinité Mts., de Granville *et al.* 6428 (P, K, U).

Vernacular names: ah-de-me-puh-tuh-puh (Tirio), put-poi-ot (Wayana; Plotkin 583); yemiki ã (Wayampi; Grenand 2103 ident. not certain).

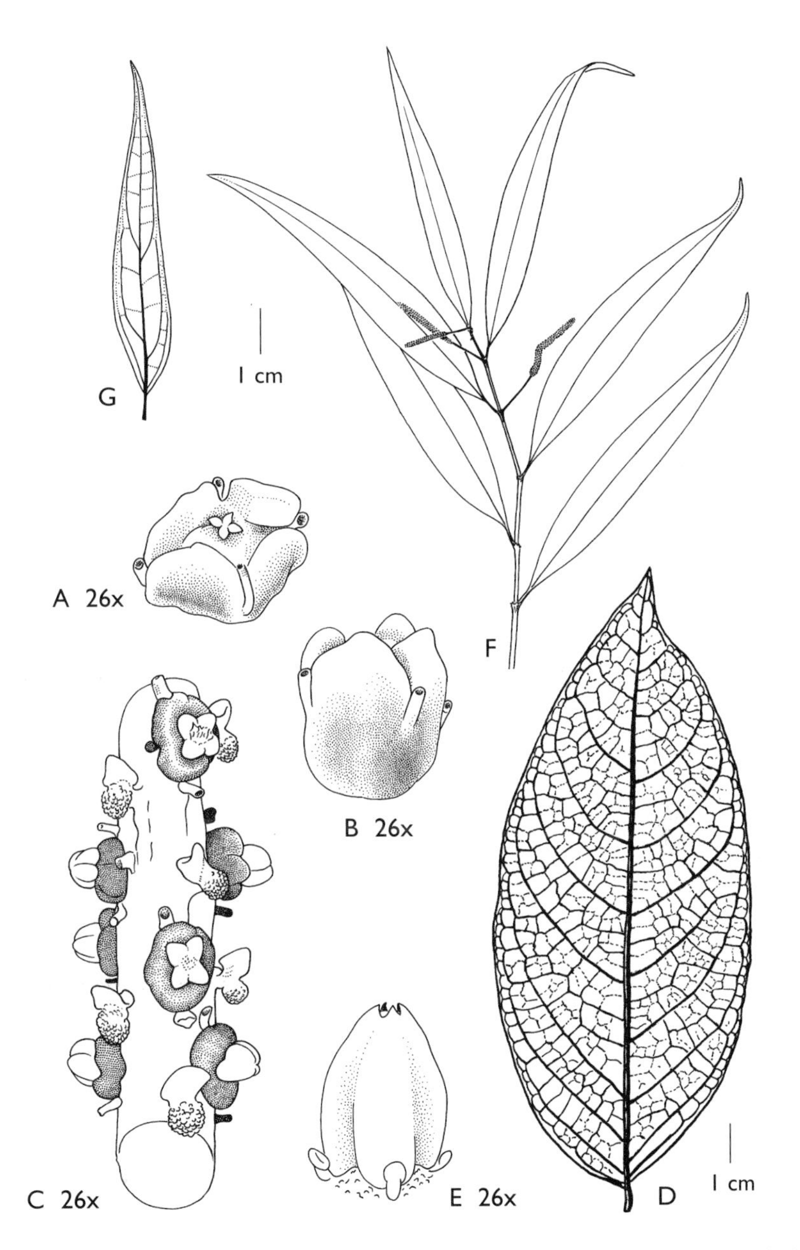

Fig. 23. *Piper alatabaccum* Trel. & Yunck.: A, fruit seen from above; B, fruit seen from the side; C, part of spike showing rachis and young flowers. *Piper bartlingianum* (Miq.) C. DC.: D, leaf; E, fruit seen from the side. *Piper bolivaranum* Yunck.: F, habit; G, leaf (A-B, Henkel 418; C, de Granville *et al.* 876; D, Kramer & Hekking 2388; E, Hoffman 1415; F-G, Maas *et al.* 5578).

Notes: The leaves have an aromatic smell.
Piper alatabaccum and *P. bartlingianum* having a rather similar appearance can easily be distinguished and recognized by the fruit characters.

5. **Piper amalago** L., Sp. Pl. 29. 1753. Type locality: Jamaica; type not designated.

Piper medium Jacq., Icon. Pl. Rar. 1: 2. 1787. – *Piper amalago* L. var. *medium* (Jacq.) Yunck., Brittonia 14: 189. 1962. – *Piper amalago* L. var. *medium* (Jacq.) Yunck. f. *medium* Steyerm., Fl. Venez. 2(2): 320. 1984. Type locality: West Indies; type not designated.
Piper ceanothifolium Kunth in Humb., Bonpl. & Kunth, Nov. Gen. Sp. ed. qu. 1: 56. 1816. – *Piper amalago* L. var. *medium* (Jacq.) Yunck. f. *ceanothifolium* (Kunth) Steyerm., Fl. Venez. 2(2): 322. 1984. Type: Venezuela, Humboldt 32 (holotype P, not seen).

Shrub, 2-4(-7) m tall. Stem glabrous or minutely pubescent on upper internodes. Prophyll covering shoot apex, 3-8 mm long. Petiole slender, 0.5-1 cm long, vaginate up to middle, glabrous or pubescent; blade lanceolate to rounded or subobovate, 8-11(-15) x 2.5-6(-10) cm, apex acuminate, base equally attached to petiole, acute to subcordate, glabrous or pubescent on veins below; palmately 5-7-veined, tertiary venation obscure. Inflorescence erect; peduncle 0.8-1.5 cm long, glabrous; spike 6-7 cm long, not apiculate; rachis minutely pubescent; floral bracts cucullate, glabrous; thecae divergent, dehiscing partially upward. Fruits ovoid, conical to apex, 1.5-2 mm long, ca. 1 mm in diam., glabrous or papillose, stigmas 3-4, sessile.

Distribution: Indo-Malayan region; West Indies, C and S America, south to Argentina and Paraguay; from sea level to 1500 m elev.; 12 collections studied (GU: 2; FG: 3).

Specimens examined: Guyana: Kanuku Mts., A.C. Smith 3458 (K); without locality, Jenman 2125 (K). French Guiana: Saül, Moretti 99, 100 (CAY), Aumeeruddy 64 (CAY).

Notes: The name amalago is from Indian origin. Nicolson, Suresh & Manilal (Regnum Veg. 119: 207-208. 1988) described its nomenclatural history. The name could be from the Malayan ammulaka or ammuluku meaning man pepper. The original illustration appears to have only male spikes. Today it is called wild pepper, kattumulaku and is used throughout Kerala, India.

Several authors discerned varieties and formas in this rather widespread and variable species. The characters used to differentiate the varieties and formas were the amount of hairs and the more or less manifest presence of a structure at the very base of the leaf. The very few collections recorded from the Guianas are identified *Piper amalago* without infraspecific entities.

6. **Piper angustifolium** Lam., Tabl. Encycl. 1: 81. 1791. Type: French Guiana, Cayenne, L.C. Richard s.n. (isotype P). – Fig. 24 A-B

Shrub or subshrub to 1.5 m tall. Stem retrorsely pubescent with swollen nodes. Stipules ovate-lanceolate, often rather long persistent; petiole 0.1-0.4 cm long, vaginate at base, retrorsely pubescent; blade not scabrous, not conspicuously glandular-dotted, narrowly lanceolate, 5-11 x 1-2.8 cm, apex long-acuminate, base equal or almost equally attached to petiole, acute, glabrous; pinnately veined, secondary veins 6-10 per side, originating from throughout primary vein, plane above, prominulous below, anastomosing well within margin. Inflorescence erect; peduncle 0.2-0.4(-0.7) cm long, retrorsely pubescent; spike almost globose, to 1 x 1.5 cm, apiculate; floral bracts cucullate, glabrous. Infructescence erect, green; fruits depressed globose or obovoid, ca. 2.5 mm wide (when dried), glabrous, stigmas 3(-4?), sessile.

Distribution: Suriname, French Guiana and Brazil (Amazonas, Rio Madeira); in primary forest, up to 500 m elev.; ca. 30 collections studied, many sterile (SU: 1; FG: 30).

Selected specimens: Suriname: s.l., UVS 17835 (BBS, U). French Guiana: Saül area, Skog & Feuillet 7271 (CAY, NY, U, US), Hahn 3633 (CAY, U, US), de Granville *et al.* 2734 (CAY, U); Sinnamary R., de Granville *et al.* 194 (CAY, P, U), Hoff 7171 (CAY, P, U); Trinité Mts., de Granville *et al.* 6219 (CAY, U); Nouragues Station, Feuillet 4417 (CAY, P, U).

Notes: In the Paris herbarium I have seen Richard s.n., an isotype. It is in a small bag and consists of two leaves, two young spikes and a stem fragment. It agrees well with the description. I have not seen the holotype and do not know where it is.
Piper angustifolium can be distinguished from *P. anonifolium* and *P. eucalyptifolium* by retrorsely pubescent stems, whereas in *P. anonifolium* and *P. eucalyptifolium* stems are glabrous, except for lines of hairs.
Comparing the descriptions and the collections at hand of *P. angustifolium* and *P. piresii* Yunck. from N Brazil I conclude that they

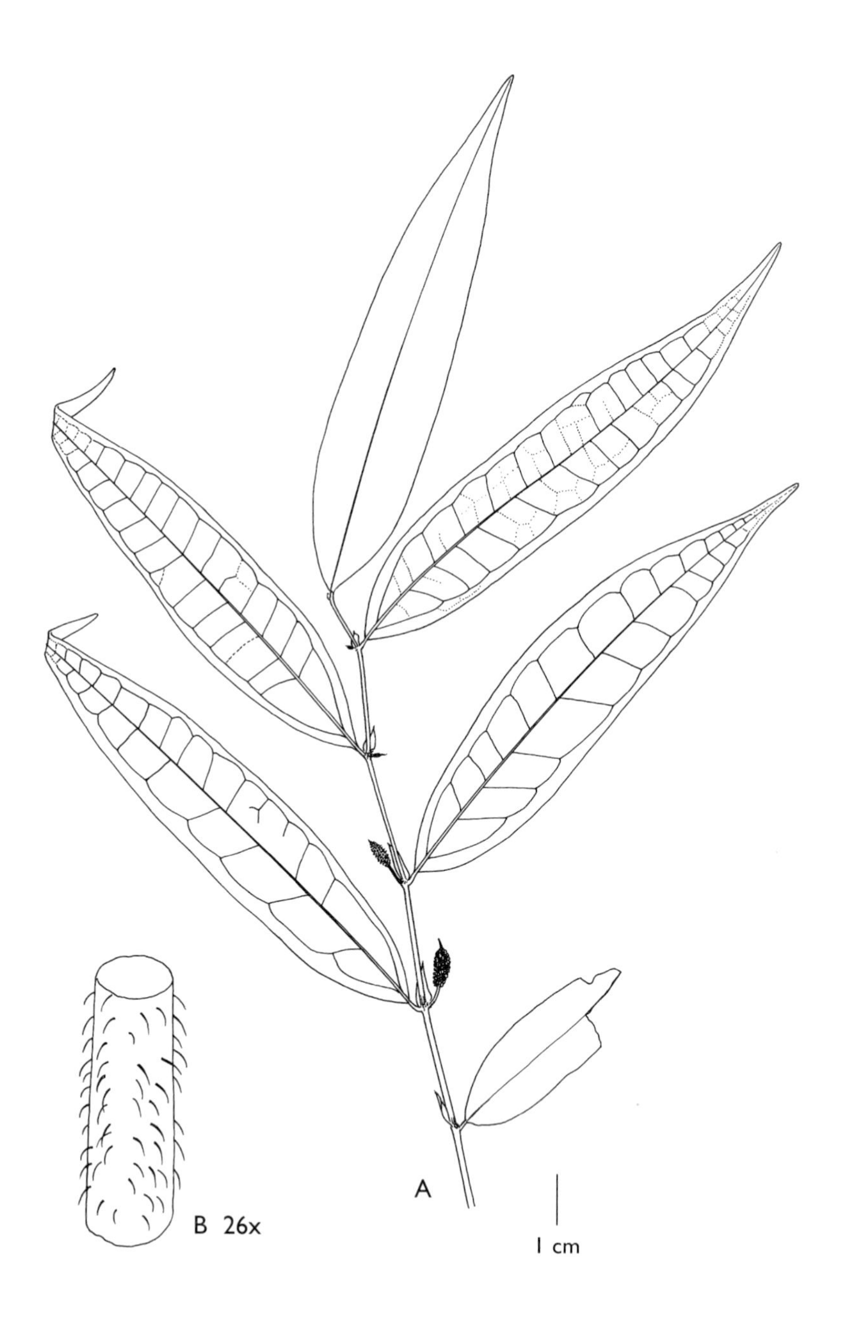

Fig. 24. *Piper angustifolium* Lam.: A, habit; B, detail of stem with indument (A-B, de Granville *et al.* 6219).

probably belong to one species. A final conclusion can only be made after studies of the type. Both are small shrubs with swollen nodes, glabrous leaves, retrorsely pubescent stems, etc. and very short spikes with cucullate, glabrous floral bracts.

7. **Piper anonifolium** (Kunth) Steud., Nomencl. Bot. ed. 2. 2: 339. 1841, as 'anonaefolium'. – *Steffensia anonifolia* Kunth, Linnaea 13: 619. 1840 ('1839'), as 'anonaefolia'. Type: French Guiana, Poiteau s.n. (holotype P, not seen).

Artanthe apiculata Klotzsch, J. Bot. (Hooker) 4: 321. 1841. Type: Guyana, Island of Coropocary on the Essequibo, Ro. Schomburgk ser. I, 53 (isotypes BM, K, US).
Piper hohenackeri C. DC. in A. DC., Prodr. 16(1): 243. 1869. Type: near Paramaribo, Kappler 1668 (holotype G-BOIS, isotype U).
Piper citrifolium sensu Trel. & Yunck. (1950: 384) and Yunck. (1957: 258) as to description and specimens cited, non Lam..
Piper anonifolium (Kunth) Steud. var. *anonifolium* f. *parvifolium* Yunck., Bol. Inst. Bot. (São Paulo) 3: 82. 1966. Type: Brazil, Pará, Peixe Boi, Huber s.n. (holotype MG, not seen).

Shrub, occasionally a tree, 0.5-3 m tall, nodose, glabrous. Stem glabrous except for 1 or 2 subnodal lines of hairs decurrent from petioles. Petiole 0.5-1 cm long, canaliculate, ciliate on margin or keel, vaginate at base only; blade thin-coriaceous to coriaceous, may be somewhat rugose, not scabrous, somewhat glandular-dotted, often bicoloured, sometimes glaucous or whitish below *in vivo,* drying silvery shiny, lanceolate to oblong to elliptic or ovate, 3-4 x longer than wide, 8-16 x 2-4.7 cm, apex acute to short-acuminate, base equal or almost equally attached to petiole, cuneate or acute, glabrous; pinnately veined, secondary veins 8-12 per side, originating from throughout primary vein, plane to impressed above, prominent below, tertiary veins reticulate. Inflorescence erect; peduncle 0.5-1 cm long; spike 2.5-4(-5) cm long, 0.2-0.5 cm thick, white, yellow or (greyish) green, apiculate; rachis glabrous; floral bracts cucullate, not fringed, often greyish when dried; anthers opening horizontally. Infructescence to 1.2(-2) cm thick; fruits depressed globose, glabrous, stigmas 3, sessile.

Distribution: The Guianas, Venezuela, Brazil (Amazonas, Rondônia, Pará, Amapá) and Amazonian Bolivia.

7a. **Piper anonifolium** (Kunth) Steud. var. **anonifolium** – Fig. 25 A-D

In the typical variety the leaves are glabrous on both sides, and at most narrowly ovate.

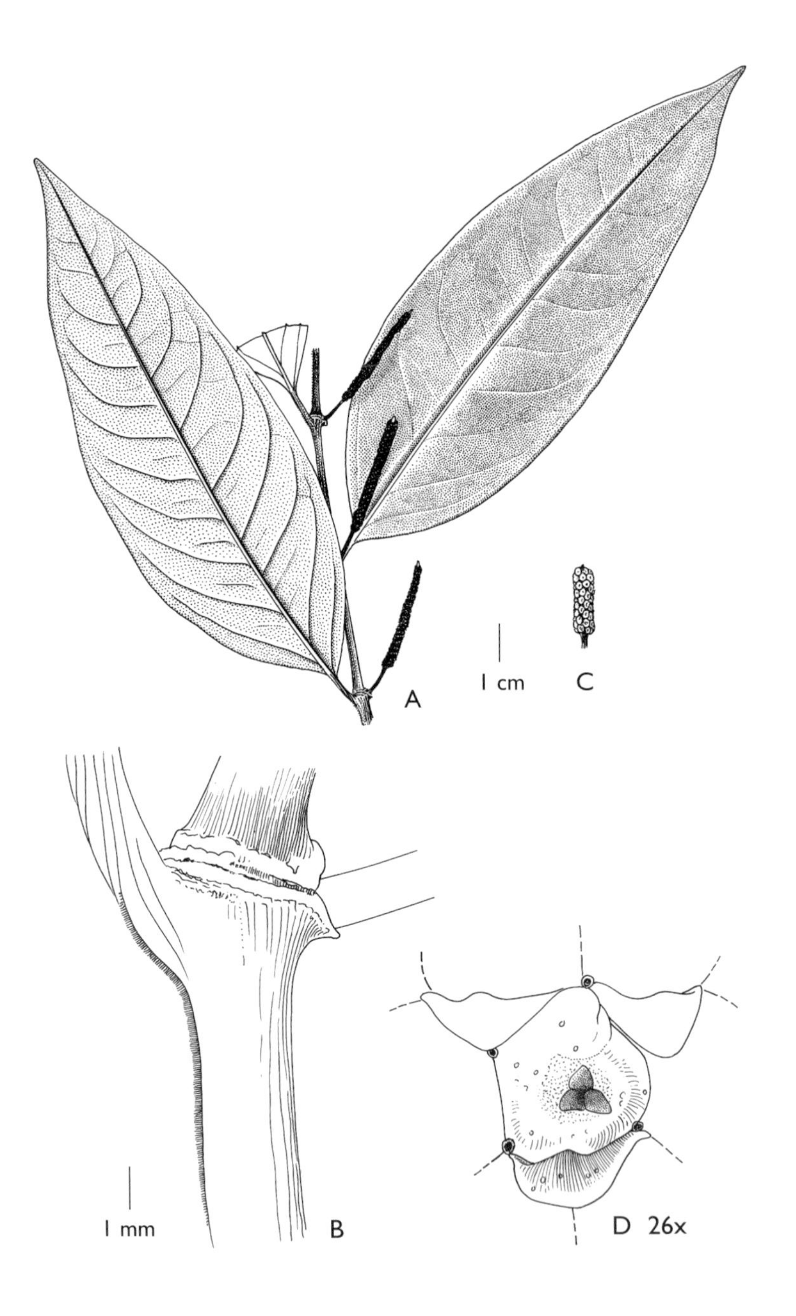

Fig. 25. *Piper anonifolium* (Kunth) Steud. var. *anonifolium*: A, habit; B, part of stem with petiole and decurrent lines of hairs; C, spike with fruits; D, fruit with stigmas seen from above (A-B, Billiet *et al.* 6313; C-D, Görts-van Rijn *et al.* 124).

Distribution: The Guianas, Venezuela, Brazil (Amazonas, Rondônia, Pará, Amapá) and Amazonian Bolivia; mainly in understory of primary forest, on various soils and granite, also in swampy forest, montane, mossy forest and occasionally in clearings; from 10-700 m elev.; over 260 collections studied (GU: 29; SU: 43; FG: 190).

Selected specimens: Guyana: Upper Takutu – Upper Essequibo region, Kassikaityu R., Clarke 4669 (U, US), Pakaraima Mts., Cashew Falls, Hoffman 1161 (U, US). Suriname: Mankaba, Sauvain 526 (BBS, CAY, P, U); between Wia wia and Grote Zwiebelzwamp, Lanjouw & Lindeman 1139 (NY, U). French Guiana: Route de Bélizon, Saül (Görts-van Rijn *et al.* 124 (CAY, NY, U); Tortue Mts., Billiet & Jadin 6313 (BR, CAY, K, MO, NY, P, U, US).

Vernacular names: Suriname: akamikini (Ndjuka; Sauvain 730), kini kini (Sauvain 526); malembe toko (Saramacca; van Donselaar 1302). French Guiana: kaboye kamwi (Palikur; de Granville *et al.* 4273). This name is not unique for the taxon.

Notes: In BM I found the collection von Rohr 232 from Cayenne bearing the name *Piper rohrii* in pencil. The collection clearly belongs to *P. anonifolium* and not to *P. rohrii* C. DC., a nom. illeg. for *P. amplum* (Kunth) Steud., which is a species from SE Brazil. Lemée (1955: 483) wrongly treated this species as occuring in French Guiana.
Piper rugosum Vahl (1797), a later homonym of *P. rugosum* Lam. (1791), described from a von Rohr collection from Cayenne, belongs to *P. anonifolium.*
I do not follow Yuncker nor Steyermark (1984: 328) in separating a f. *parvifolium*. The species is quite variable in leaf shape and there is no reason to recognize a separate forma.
I agree with Steyermark (1984: 324) that *P. citrifolium* as described by Trelease and Yuncker (1950: 384) is conspecific with *P. anonifolium.* In the protologue Lamarck described *P. citrifolium* as being hirtellous. This description is very brief. The description by Kunth (1840: 629) - based on Lamarck's – is more elaborate. The taxon thus described does not agree with *P. anonifolium.*
Piper eucalyptifolium Rudge is morphologically rather similar to *P. anonifolium*; both taxa are glabrous except for lines of hairs on petiole, decurrent on internodes. Both have short spikes with glabrous, cucullate floral bracts. For the time being the two can better be kept apart. They can be differentiated by the shape of the leaves: length/width ratio 4-6 times longer than wide and apex long-acuminate characterize *P. eucalyptifolium*, whereas in *P. anonifolium* the ratio is 3-4 times longer than wide, and the leaf apex is acute to shortly acuminate. If further

studies - especially on inflorescence and fruit structures - conclude that there is only one taxon, then Rudge's name has priority.

7b. **Piper anonifolium** (Kunth) Steud. var. **parkerianum** (Miq.) Steyerm., Fl. Venez. 2(2): 328. 1984. – *Artanthe parkeriana* Miq., London J. Bot. 4: 467. 1845. – *Piper parkerianum* (Miq.) C. DC. in A. DC., Prodr. 16(1): 373. 1869. – *Piper citrifolium* Lam. var. *parkerianum* (Miq.) Trel. & Yunck., Piperac. N. South Amer. 385. 1950. Type: Guyana, Demerara, Parker s.n. (holotype K).

Piper brachystachyum C. DC. in A. DC., Prodr. 16(1): 296. 1869, non *Piper brachystachyon* Vahl 1804. – *Piper miquelii* C. DC, Candollea 1: 238, 261. 1923.Type: French Guiana, near Karouany, Sagot 542 (lectotype K, isolectotypes BM, P) (designated by Trel. & Yunck. 1950: 384).

This variety differs in the indument on the leaves. In the typical variety the leaves are glabrous on both sides, whereas in var. *parkerianum* the leaves are puberulent below. The shape of the leaves is not narrowly ovate, but ovate.

Distribution: Venezuela, Guyana, French Guiana: 19 collections studied (GU: 18; FG: 1).

Selected specimens: Guyana: Barima-Waini region, head of Barima R., Ayanganna Falls, Pipoly 8204 (U, US); Cuyuni-Mazaruni region, Kangu R., near E peak of Mt. Ayanganna, Pipoly 11051 (U, US); Potaro-Siparuni region, North Fork R., McDowell 4855 (U, US); Essequibo R., near Rockstone, Maas 3935 (BRG, U); Essequibo R., Groete Cr., Maguire *et al.* 22864 (U); N facing slope of Mt. Roraima, K.E.R. Mt. Roraima expedition, Warrington 279 (K).

Note: Trelease & Yuncker (1950: 384) described *Piper miquelii* as having subnodal lines of hairs and the leaf lower surface pubescent including the veins. The type specimens at BM and K show the characteristic subnodal lines in a very few places. Presumably the lines disappear on older branches. The specimens also show the characteristic short, thick, apiculate spikes with cucullate, often greyish floral bracts. As a result, it is appropriate to assign Sagot 542, the type of *P. miquelii*, to *P. anonifolium* var. *parkerianum*.

8. **Piper arboreum** Aubl., Hist. Pl. Guiane 1: 23. 1775. – *Piper verrucosum* Sw., Prodr. 15. 1788, nom. illeg. – *Artanthe lessertiana* Miq., Syst. Piperac. 405. 1844, nom. illeg. Type: French Guiana, Aublet s.n. (holotype P, not seen, isotype BM, BM photo 3472).
– Fig. 26 A-F

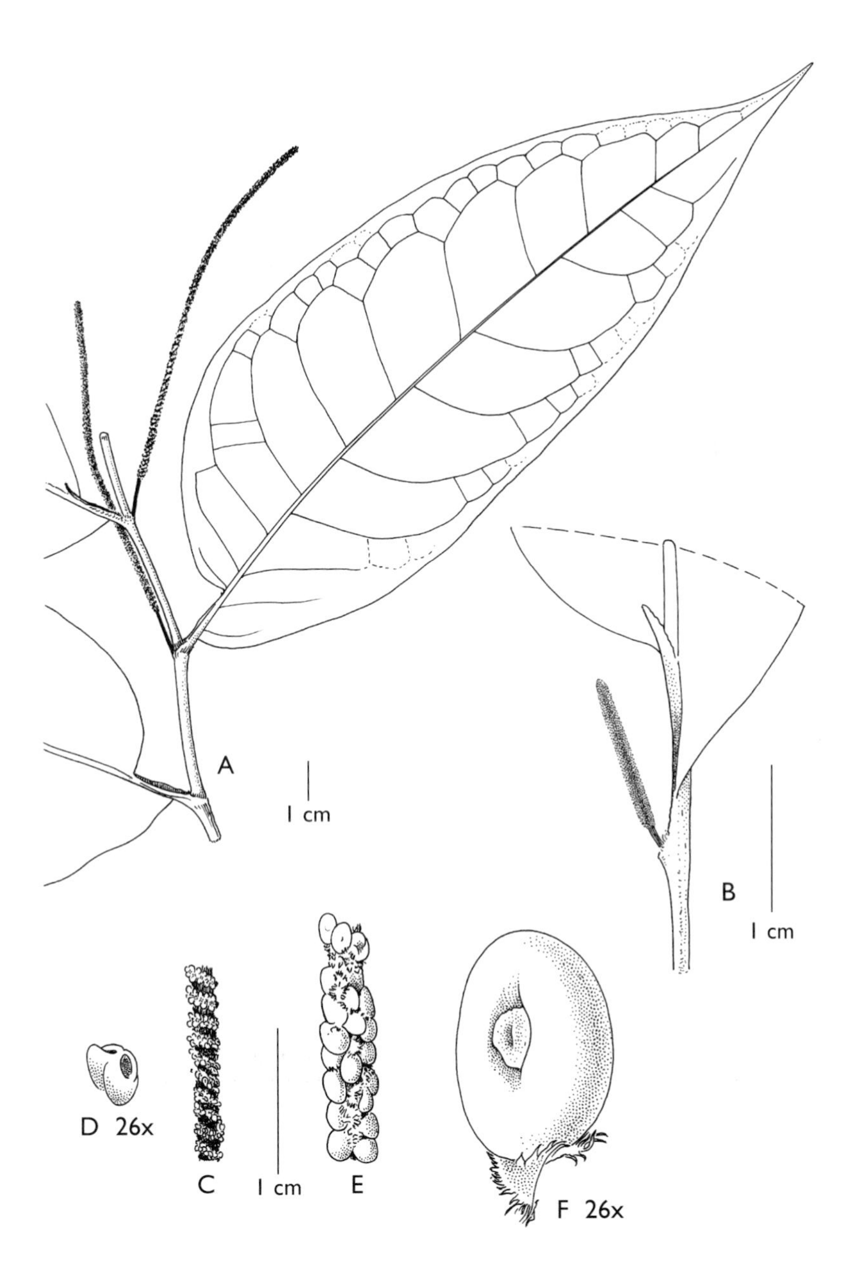

Fig. 26. *Piper arboreum* Aubl.: A, habit; B, tip of branch, young spike, prophyll and leaf attachment; C, detail of spike showing spirally arranged flowers; D, anther; E, detail of spike with fruits; F, fruit with subtending floral bract (A-D, Hahn 5533; E-F, Fanshawe 3076).

Piper geniculatum Sw., Prodr. 15. 1788. – *Steffensia geniculata* (Sw.) Kunth, Linnaea 13: 612. 1840 ('1839'). Type: Jamaica, Swartz s.n. (holotype G).
Piper falcifolium Trel., Contr. U.S. Natl. Herb. 26: 25. 1927. – *Piper arboreum* Aubl. var. *falcifolium* (Trel.) Yunck., Ann. Missouri Bot. Gard. 37: 65. 1950. Type: Panama, Chiriquí, San Felix, Pittier 5137 US (holotype US, not seen).
Piper arboreum Aubl. var. *hirtellum* Yunck., Ann. Missouri Bot. Gard. 37: 64. 1950. Type: Guyana, NW Distr., Waini R., de la Cruz 3623 (holotype US).
Piper arboreum Aubl. var. *paucinervium* Trel. & Yunck., Piperac. N. South Amer. 374. 1950. Type: French Guiana, Cayenne, Broadway 854 (holotype US, isotypes GH, NY, only seen K).

Shrub or treelet, 2-7 m tall, sometimes with long, arching branches, often strongly branching. Stem glabrous or minutely pubescent, upper internodes smooth or somewhat lenticellate. Shoot apex emerging from within sheathing leaf base, prophyll undeveloped. Petiole glabrous or minutely pubescent, 0.3-2 cm long, deeply vaginate, margin protracted beyond base of blade forming a ligule-like structure to 4 mm long; blade shiny green above, lighter to whitish below, oblong-ovate, (narrowly) ovate or (narrowly) elliptic, 10-25 x 6-11 cm, apex acute to (long-)acuminate, base unequally attached to petiole difference 0.5-1(-3) cm (less in young leaves), longest side obtuse, rounded to cordulate, in narrow leaves acute, shorter side acute to obtuse, glabrous or minutely pubescent below; pinnately veined, secondary veins 8-10 per side, originating from throughout primary vein, loop-connected towards apex, tertiary veins inconspicuously reticulate. Inflorescence erect; peduncle moderately stout, 0.5-1 cm long; spike 5-10(-17) cm long, whitish, yellow or pale green, apiculate, densely flowered; rachis glabrous; floral bracts 0.3-0.6 mm in diam., densely marginally fringed, conspicuously arranged in whorls; anthers dehiscing apically. Fruits dark green, laterally compressed, oblongoid or obovoid, 2 mm long, 0.8-1 mm in diam., glabrous or slightly papillose, stigmas 3-4, sessile.

Distribution: Mexico south to S Brazil and Paraguay, West Indies; in moist tropical rain forest high, dense, mixed or swamp forest, occasionally in secondary vegetation, from sea level to 1500 m elev.; ca. 240 collections studied (GU: 94; SU: 86; FG: 61).

Selected specimens: Guyana: Rupununi Distr., Kuyuwini Landing, Kuyuwini R., Jansen-Jacobs *et al.* 2337 (BBS, BRG, NY, P, U); Kanuku Mts., Rupununi R., Puwib R., Jansen-Jacobs *et al.* 301 (BBS, BRG, K, NY, U). Suriname: Kabalebo dam area, Lindeman & Görts-van Rijn *et al.* 140 (BBS, K, U); Piki pada, Sauvain 726 (CAY, BBS, U, US). French Guiana: Saül, de Granville *et al.* 2345 (CAY, NY, P, U, US); Upper Camopi R., Mt. Belvédère, de Granville *et al.* 6957 (CAY, NY, P, U).

Vernacular names: Guyana: wild thick leaf (Reinders 165); Suriname: gewone malimbe-toko (Daniëls & Jonker 1072); mukaj katenga = akajatenga = apuku pepe = kolakatenga (Boni; Sauvain 122); kakakatenga (Boni; Sauvain 336); kakakaringa (Sauvain 628); abaon dèku (Saramacca; Sauvain 726). French Guiana: yak takwã (Wayampi; Haxaire 568); gama na udu (Ndjuka; Daniel 3).

Use: According to Reinders, *Piper arboreum* is used against poisonous snake bites in the NW Distr., Guyana.

Notes: One of the most common species; recognizable by the (almost) glabrous habit and the very large difference of blade attachment at the petiole.
Tebbs (1989: 156) considered the BM specimen of Aublet s.n. from French Guiana to be the holotype. In their article "Un nouvel herbier de Fusée Aublet découvert en France" (Recueil Trav. Bot. Néerl. 37:133-170. 1940) Lanjouw & Uittien published a list of Aublet collections present in herb. Denaiffe. *Piper arboreum* is nr. 68 in their list and is referred to herb. Denaiffe Vol. 1, nr 36. They noted: "Des feuilles seulement", thus a sterile collection. The specimens in herb. Denaiffe (which now is in P) have labels in Aublet's handwriting, whereas those in BM have not (Lanjouw & Uittien, p. 145). The Denaiffe specimen (that I could not find) thus is the holotype, and the BM specimen is an isotype.
There is a collection Ro. Schomburgk ser. II, 903 from Roraima in BM which is somewhat atypical in having large, crushed fruits and rather small leaves. It shows, however, the typical glabrous leaves with unequal base.
Trelease & Yuncker created *P. arboreum* var. *paucinervium* based on two French Guianan collections by Broadway 759 and 854 made near Cayenne. This variety was created to describe narrow-leaved specimens. *Piper arboreum* is a very variable taxon and var. *paucinervium* seems to fit well within the typical variety.
Lemée (1955: 484) mentioned *P. lapathifolium* (Kunth) Steud. for French Guiana, referring to C. DC (1869: 323) who for this species listed a French Guianan specimen, Leprieur 145 in G-DEL. This collection clearly belongs to *P. arboreum*. Miquel (1844: 405) listed this specimen under his *Artanthe lessertiana*, nom. illeg. for *P. verruccosum* Sw., which in turn is a nom. illeg. for *P. arboreum*. *Piper lapathifolium* occurs in Mexico.
For differentiating characters with *P. tuberculatum* see note to the latter.

9. **Piper augustum** Rudge, Pl. Guian. Rar. 10, t. 7. 1805. Type: French Guiana, J. Martin s.n. (holotype BM). – Fig. 27 A-C

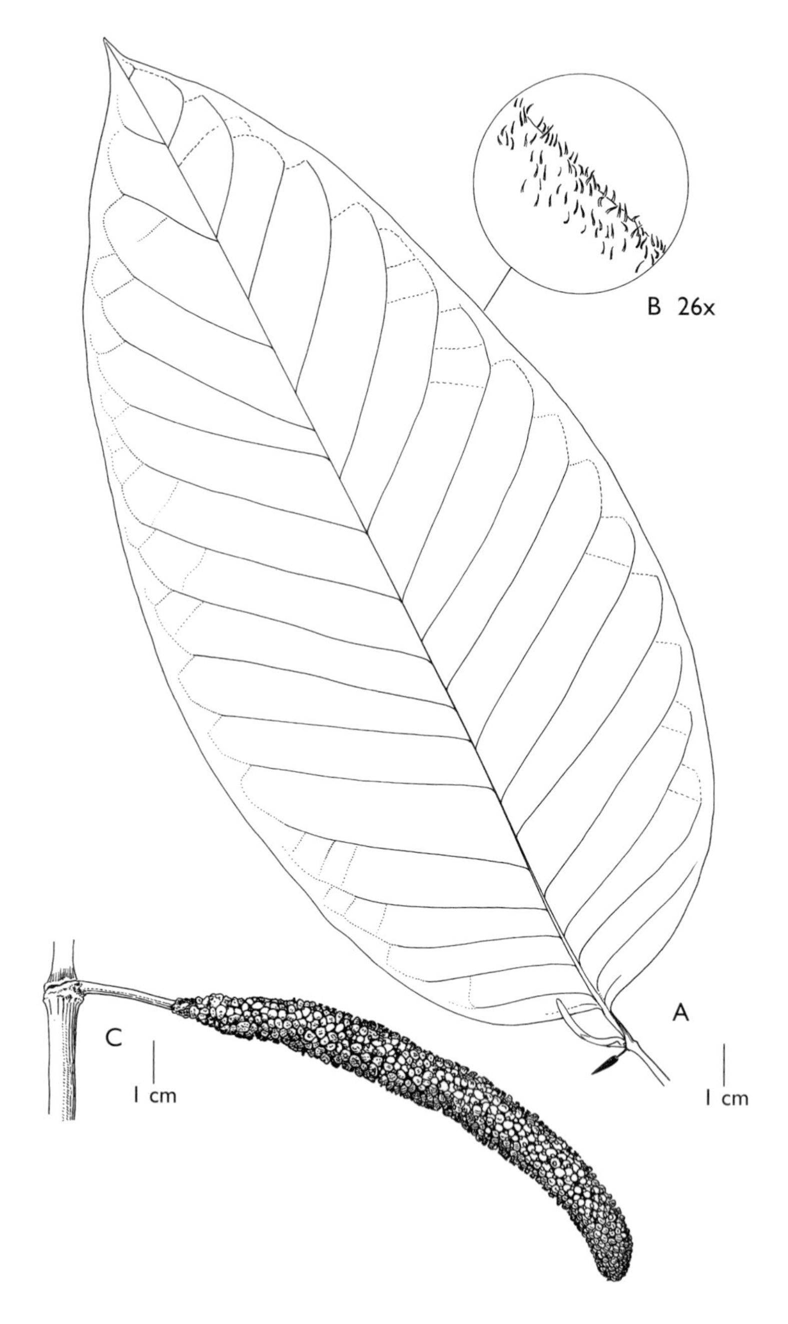

Fig. 27. *Piper augustum* Rudge: A, habit; B, detail of leaf margin; C, spike with fruits (A-B, de Granville *et al.* 7992; C, McDowell 4748).

Shrub or treelet, to 2-4(-8) m tall, glabrous, except for ciliate leaf margin. Petiole 0.5-4 cm long, vaginate at base; blade glandular-dotted, elliptic-ovate to oblong-elliptic, 15-35 x (6-)9-15(-18) cm, margin ciliate, apex acute, base almost equal to unequally attached to petiole, acute or obtuse, occasionally truncate; pinnately veined, secondary veins 10-16 per side, strongly curved, originating from throughout primary vein. Inflorescence pendent, may be erect or horizontal at first; peduncle glandular-dotted, 1-2(-3) cm long, red to brown; spike 4-20 cm long, to 2.5 cm diam. (*in vivo),* (greenish) white or green with purple points, not or slightly apiculate; rachis glabrous, sometimes red; floral bracts densely marginally fringed. Fruits oblongoid, glabrous, may be somewhat papillose, to 4 mm long, 2 mm in diam., becoming stylose or mammiform at apex, stigmas obsolete.

Distribution: From Costa Rica to northern S America and Peru; often in humid places, also recorded from secondary vegetation, from sea level to 1500 m elev.; ca. 90 collections studied (GU: 12; SU: 6; FG: 70).

Selected specimens: Guyana: Pakaraima Mts., Paruima, Maas *et al.* 5601 (BRG, U); Cuyuni-Mazaruni region, McDowell 4748 (U, US). Suriname: Oelemari R., Wessels Boer 1095 (NY, U); Tumac Humac Mts., Acevedo 5987 (U, US). French Guiana: Saül, Boom 10787 (NY, U); Görts *et al.* 108 (CAY, NY, U).

Uses: Leaves have a spicy odour when crushed. In Amazonia Indians use stems as tooth brush. It was also used to prevent tooth decay (Davis & Yost, Bot. Mus. Leafl. 29(3): 182. 1983).

Vernacular name: French Guiana: gan bushi man lembe lembe (Boni; Fleury 178).

Notes: *Piper augustum* is easily recognizable by the large number of secondary veins, the ciliate leaf margin and the thick infructescence.
Yuncker (1957: 265) recognized *P. augustum* var. *pubinerve* Trel. & Yunck. with the veins beneath puberulent, said to occur in Guyana and French Guiana, and expected to occur in Suriname. I have no indication that this taxon, originally described from Chocó in Colombia, occurs in the Guianas.

10. **Piper aulacospermum** Callejas in Görts & Callejas, Blumea 50: 367. 2005. Type: French Guiana, near Saül, la Fumée Mt., Mori *et al.* 18087 (holotype NY, isotypes CAY, HUA). – Fig. 28 A-H

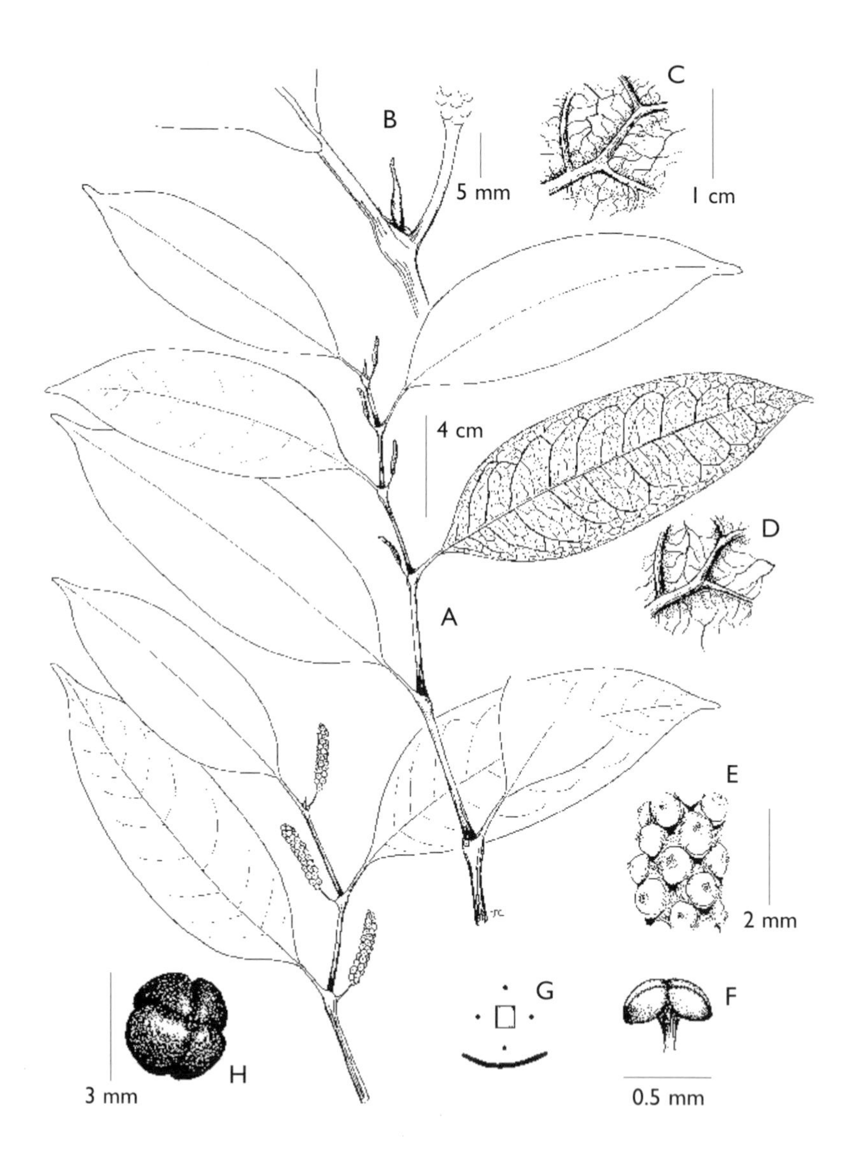

Fig. 28. *Piper aulacospermum* Callejas: A, habit, showing sympodial branches with inflorescences at anthesis and fruit; B, detail of leaf base and prophyll; C, leaf veins, view from above; D, leaf veins, view from below; E, young fruits; F, anther; G, floral diagram; H, seed (A-H, Mori & Pennington 18087). Drawing by J. Colorado (HUA).

Small shrub, 1-2.5 m tall, sparingly branched. Stem striate or canaliculate, glabrous. Prophyll 2-4 cm long, glabrous, caducous. Petiole 0.7-1.3 cm long, glabrous, deeply vaginate at base; blade thick chartaceous, not scabrous, not conspicuously glandular-dotted, oblong or elliptic-oblong, 13-16.5 x (3.5-)4-6.5 cm, apex abruptly long-acuminate, base equal or almost equally attached to petiole, acute, glabrous; pinnately veined, secondary veins 8-12 per side, prominent on both sides, anastomosing well within margin, originating from throughout primary vein, tertiary veins widely, distinctly reticulate. Inflorescence erect, white, becoming green; peduncle 1 cm long, minutely pubescent; spike 0.5-1 x 0.1-0.2 cm long, in fruit 2.5-4.5 x 0.3-0.45 cm, not apiculate; floral bracts saccate or galeate, glabrous, densely fimbriate; flowers densely congested; anthers reniform. Fruits 4-sulcate, globose, 3-4.5 mm in diam., truncate at apex, glabrous, stigmas 4, globose, sessile; seeds deeply 4-sulcate.

Distribution: The Guianas; in lowland moist forest, 200-450 m elev. (GU: 1; SU: 1; FG: 10).

Specimens examined (by Callejas): French Guiana, La Fumée Mt., Marshall & Rombold 138 (NY), Mori *et al.* 14919 (NY), 15399 (NY), 15491 (HUA, NY).

Note: Several collections recorded from French Guiana representing *Piper aulacospermum* had been assigned to *P. piscatorum*. Comparing the type of the latter and collections used to describe *P. aulacospermum* it became obvious that the French Guianan collections belong to *P. aulacospermum*. Differentiating characters are: in *P. aulacospermum* the secondary veins prominent above, tertiary venation distinctly reticulate (instead of impressed or slightly prominulous in *P. piscatorum*); peduncle 1 cm long, slightly pubescent (instead of 0.4-0.5(-1) cm long, glabrous in *P. piscatorum*); the fruits are 4-sulcate (caused by the shape of the seeds), 3-4.5 mm in diam. (instead of smooth, ca. 1.5 mm in diam. in *P. piscatorum*).

11. **Piper avellanum** (Miq.) C. DC. in A. DC., Prodr. 16(1): 302. 1869. – *Artanthe avellana* Miq., Syst. Piperac. 478. 1844. Type: French Guiana, Leprieur s.n. (holotype G-DEL, isotype U). – Fig. 29 A-B

Artanthe gabrieliana Miq., Syst. Piperac. 476. 1844. – *Piper gabrielianum* (Miq.) C. DC. in A. DC., Prodr. 16(1): 268. 1869, as 'gabrielanum'. Type: French Guiana, Gabriel s.n. (holotype G-DEL), syn. nov.
Piper marowynense C. DC. in A. DC., Prodr. 16(1): 268. 1869. Type: Suriname, Marowijne R., Monte Cassinisoord, Wullschlaegel 1562 (holotype BR, not seen).

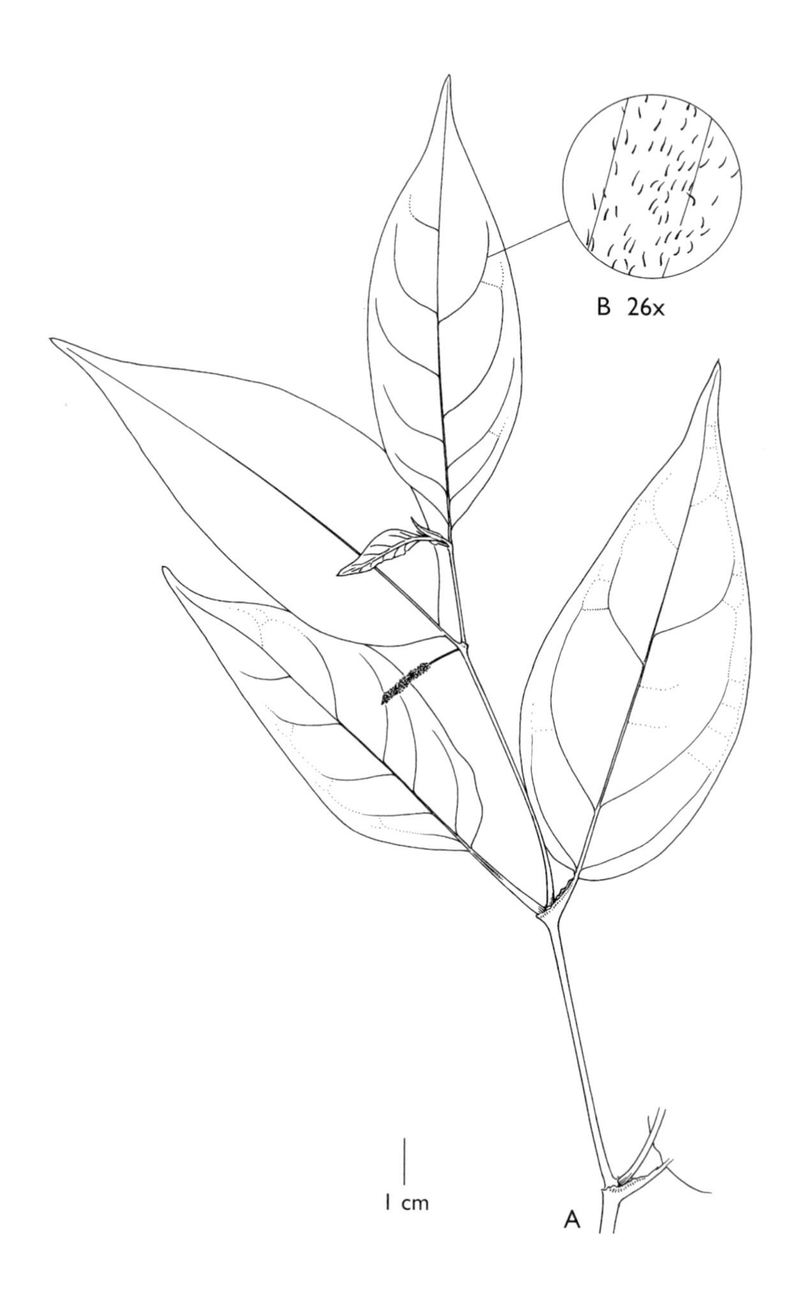

Fig. 29. *Piper avellanum* (Miq.) C. DC.: A, habit; B, detail of vein with indument (A-B, McDowell 3735).

Piper acarouanyanum C. DC. in A. DC., Prodr. 16(1): 311. 1869. Type: Acarouany, near Cayenne, Sagot 541 (holotype P, isotype K).
Piper moraballianum Trel., Bull. Misc. Inform. Kew 1933: 338. 1933. Type: Guyana, Essequibo R., Moraballi Cr., Sandwith 361 (holotype and isotype K).

Shrub or subshrub to 2 m tall, densely short crisp-pubescent, hairs to 0.2 mm long. Petiole 0.3-1.3 cm long, vaginate to apex, often conspicuously winged, green or maroon; blade not conspicuously glandular-dotted, lance-elliptic or lanceolate-oblong or oblanceolate, 7-18 x 3-7.5 cm, margin ciliate, apex short-acuminate or acute, base unequally attached to petiole difference 0.1-0.5 cm, rounded, obtuse or subacute, glabrous or sparsely short-pubescent above, sparsely to densely so below; pinnately veined, secondary veins 5-6(-8) per side, originating from lower 3/4 of primary vein. Inflorescence pendant; peduncle slender, 0.5-1 cm long, pubescent; spike 3-5.5(-10) cm long, white, yellow or green, occasionally apiculate or not; rachis glabrous; floral bracts densely marginally fringed. Fruits trigonous or depressed globose, to 2 mm in diam., glabrous and papillose, stigmas 3, sessile.

Distribution: Venezuela and the Guianas; in (dense) forest, in river-bank or marsh forest and in secondary vegetation; 90 collections studied (GU: 31; SU: 20; FG: 40).

Selected specimens: Guyana: Potaro-Siparuni region, Iwokrama Mts., Iwokrama reserve, Mutchnick 900 (U); Rupununi Distr., Kuyuwini R., Kuyuwini Landing, Jansen-Jacobs *et al.* 2923 (BRG, U). Suriname: near Paramaribo, Hekking 874 (U); near LBB post Perica along road Meerzorg-Albina, Heyde & Lindeman 276 (U). French Guiana: Lac des Americains, Ile de Cayenne, de Granville *et al.* 9127 (B, CAY, NY, P, U, US); Chemin de croix de Bourda, Jacquemin 1986 (CAY, U).

Vernacular name: Guyana: warakaba bush (Creole; van Andel 1800).

Use: Sap of crushed leaves is said to be used against snake-bites in the NW district of Guyana (Van Andel, pers. comm.).

Notes: For distinguishing characters with *Piper adenandrum* see note to the latter.
Piper rugosum Lam., Tabl. Encycl. 1: 81. 1791 was based on two specimens, one of which originated from French Guiana, the other from S Domingo. The French Guiana specimen, collected in Cayenne by Stoupy, has not been located. Even though Lemée (1955: 479) accepted it as the correct name of a French Guianan species, based on the poor description we assume that it fits within *P. avellanum*.

Yuncker (1957: 246) mentioned *P. avellanum* (Miq.) C. DC. var. *angustifolium* Trel. & Yunck., based on a Suriname collection by Weigelt s.n. (holotype PH). I suppose that it fits within the typical variety, but as I have not seen the collection I cannot add it to the synonyms. No further collections have been reported from the Guianas.
Piper gabrielianum, known only from the type collection, seems to be conspecific with *P. avellanum.* The specimen has the same kind of indument and other leaf characters. The spikes are too young to measure fruits. The lack of this character makes it difficult to place the taxon, but certainly does not support maintaining it separate.
See under *P. paramaribense* for a note on *P. pertinax.*

12. **Piper bartlingianum** (Miq.) C. DC. in A. DC., Prodr. 16(1): 257. 1869. – *Artanthe bartlingiana* Miq., Syst. Piperac. 510. 1844. Type: French Guiana, Poiteau s.n. (holotype G-DEL, isotypes 2xK?, U). – Fig. 23 D-E

Artanthe warakabacoura Miq., London J. Bot. 4: 469. 1845. – *Piper warakabacoura* (Miq.) C. DC. in A. DC., Prodr. 16(1): 257. 1869, as 'warakaboura'. Type: Guyana, Parker s.n. (holotype K, isotype U).

Shrub, subshrub or treelet, 1-4 m tall, somewhat nodose, glabrous. Petiole 0.5 cm long, glabrous, vaginate near or at base; blade glossy, coriaceous, not scabrous, not conspicuously glandular-dotted, elliptic-oblong, 13-30 x 5-8(-10) cm, apex acuminate, base equal or almost equally attached to petiole, acute, glabrous; pinnately veined, secondary veins 6-8 per side, flat to impressed above, prominent below, anastomosing well within margin, originating from throughout primary vein, tertiary veins widely reticulate. Inflorescence erect; peduncle 1 cm long, glabrous; spike 10-12 cm long, green, not apiculate; rachis hirsute; floral bracts cucullate, pilose on inner side. Fruits ovoid to tetragonous, separate, somewhat ridged or winged (at least when dried), glabrous may be somewhat papillose, stigmas 4, sessile.

Distribution: The Guianas, S Venezuela, Brazil (Amazonas, Roraima, Amapá, Pará, Matto Grosso) and SE Colombia; in forest on terra firme and in marsh forest, on sand, clay or granite, up to 700 m elev.; ca. 185 collections studied (GU: 50; SU: 50; FG: 87).

Selected specimens: Guyana: Marudi Mts., Stoffers & Görts-van Rijn *et al.* 335 (BRG, NY, U, US); Kamoa R., Jansen-Jacobs *et al.* 1593, (BRG, K, NY, P, U). Suriname: S of Juliana peak, N of Lucie R., Irwin 57708 (K, NY, U); Ulemari R., UVS 17751 (BBS, U). French Guiana: Saül area, Mt. Galbao, de Granville *et al.* 8545 (B, BR, CAY, G, INPA, MG, MO, NY, P, U, US); Saül, Görts-van Rijn *et al.* 65 (CAY, NY, U, US).

Vernacular names: Suriname: man-aneise-wiwiri (Arowak); akamikini (Saramacca; Sauvain 286), petpe (Wayana; Stahel 67). French Guiana: petpe (Wayana; Fleury 980, 1973); yemilã (Wayampi; Jacquemin 1709), kaboye (Palikur; Grenand 2121); pao pao (Garnier 90). These names are not unique for the taxon.

Use: According to Miquel, *Artanthe warakabacoura* – meaning knee of the Warakaba bird – was used in Guyana and Suriname as one of the constituents of the Ourali poison.

Notes: I agree with Trelease and Yuncker (1950: 410) that *Piper warakabacoura* and *P. bartlingianum* are conspecific. The differences used to separate the two, for instance by De Candolle, are on leaf size and shape and a few hairs on the fruits. With the large amount of collections studied, it can be concluded that the variation is such that it is not possible to maintain two species.
See also note to *P. alatabaccum*.

13. **Piper bolivaranum** Yunck., Fieldiana Bot. 28: 205. 1951. Type: Venezuela, Bolívar, Ptari-tepuí, Steyermark 59468 (holotype F).
– Fig. 23 F-G

Subshrub to 0.1-1.5 m tall. Stem glabrous. Petiole (0.2-)0.4-0.6(-1) cm long, glabrous, vaginate to apex; blade not scabrous, not conspicuously glandular-dotted, narrowly lanceolate, 7-10.5(-16.5) x 0.7-1.5(-4.5) cm, apex long-acuminate, base equal, acute, glabrous above, glabrescent or glabrous below; pinnately veined, secondary veins 2-3 per side, originating from lower 1/2 or 2/3 of primary vein, with many intermediates up to apex, secondary veins plane above, prominulous below, tertiary veins parallel. Inflorescence erect or pendent; peduncle 0.5-1.1(-2.5) cm long, glabrous; spike1-1.7(-4.5) cm long, white, not apiculate; floral bracts trigonous, densely marginally fringed. Infructescence erect, green; fruits trigonous, to 1 mm wide, glabrous, green, stigmas 3, sessile.

Distribution: S Venezuela and Guyana; 800-1100 m elev.; 2 collections studied (GU: 2).

Specimens examined: Guyana: Cuyuni-Mazaruni region, Paruima, Clarke 5476 (U, US); Pakaraima Mts., Kamarang, Maas *et al.* 5578 (BRG, U).

Note: See note to *P. fanshawei*.

14. **Piper brasiliense** C. DC. in A. DC., Prodr. 16(1): 259. 1869. – *Peltobryon pubescens* Miq., Linnaea 20: 134. 1847 (non *Piper pubescens* Vahl 1804). Type: E Colombia (formerly Brazil), Upper Japurá, Rio Negro, Martius s.n. (holotype M, not seen, isotype U). – Fig. 30 A-B

Piper liesneri Steyerm., Fl. Venez. 2(2): 467, f. 67. 1984. Type: Venezuela, Amazonas, 4 km NE of San Carlos de Rio Negro, Liesner 6405 (holotype VEN, isotype MO, none seen), syn. nov.

Small shrub, to 1 m tall. Stem densely retrorsely pubescent with hairs to 1 mm long. Petiole 0.5-1 cm long, densely pubescent with long hairs, vaginate at base; blade not scabrous, densely glandular-dotted below, (narrowly) elliptic to subobovate, 9.5-16 x 3.5-7.2 cm, apex acute to slightly acuminate, base unequally attached to petiole difference 0-0.2 cm, narrowing to lobed base, lobes shorter than petiole, unequal, longer one subreniform, overlapping petiole, blade glabrous above, densely pubescent with long hairs below including veins; pinnately veined, secondary veins 6-11 per side, originating from throughout primary vein, prominulous above, prominulous below, tertiary veins reticulate. Inflorescence pendent; peduncle 0.3-0.4 (0.8 in fruit) cm long, densely pubescent; spike 1-1.5 cm long, 0.5 cm thick, not apiculate?; floral bracts narrowly sublunulate, sparsely puberulous. Infructescence pendent, 2.5 cm long, ca. 1 cm thick; fruits globose, depressed at apex, densely crisp-pubescent, stigmas 3, recurved, on long, pilose style.

Distribution: Colombia, Venezuela and Guyana, Peru; up to 650 m elev.; 4 collections studied (GU: 3).

Specimens examined: Guyana: Potaro-Siparuni region, Clarke 8916 (U); Upper Takutu-Upper Essequibo region, Chodikar R., Clarke 2805 (U); Mt. Ayanganna, Tillett *et al.* 45177 (NY).

15. **Piper brownsbergense** Yunck. in Trel. & Yunck., Piperac. N. South Amer. 37. 1950. Type: Suriname, Brownsberg, BW 3172 (holotype U). – Fig. 31 A-B

Small shrub or subshrub, 0.15-1.5 m tall. Stem somewhat yellow-glandular, pilose, hairs to 2 mm long; upper internodes drying finely ridged. Petiole (0.2-)0.3-0.5(-3.5) cm long, pilose; blade glandular-dotted, somewhat scabrous, elliptic, lanceolate-oblong or ovate, (7-)11-21 x (3-)4-11 cm, margin ciliate, apex acute to acuminate, base equal or unequally attached to petiole difference 0.2-0.3 cm, subcordate or obtuse, shorter side cuneate, pilose above, more densely so below;

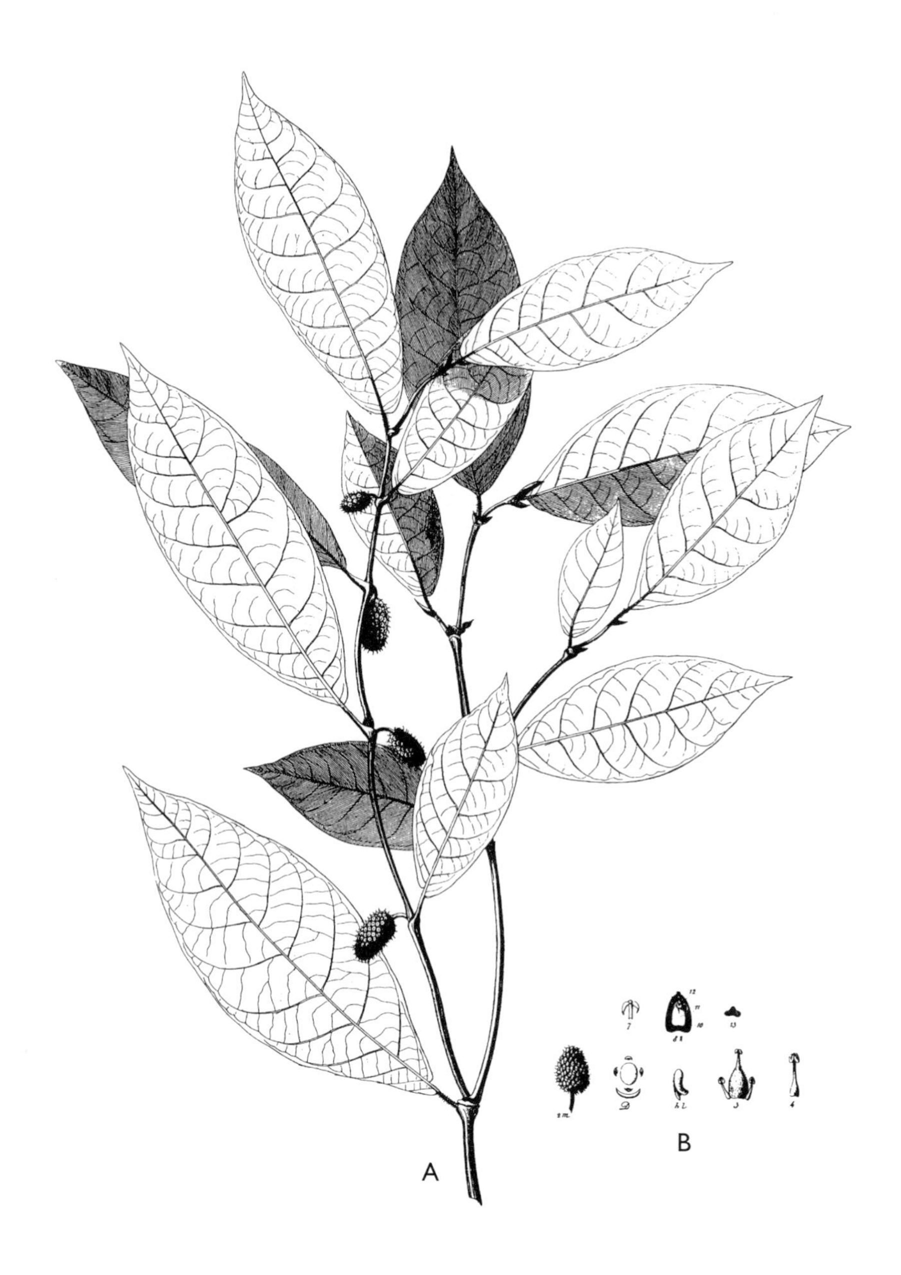

Fig. 30. *Piper brasiliense* C. DC.: A, habit; B, flower details (A-B, after illustration in Fl. Bras. 4(1), fig. 4. 1852).

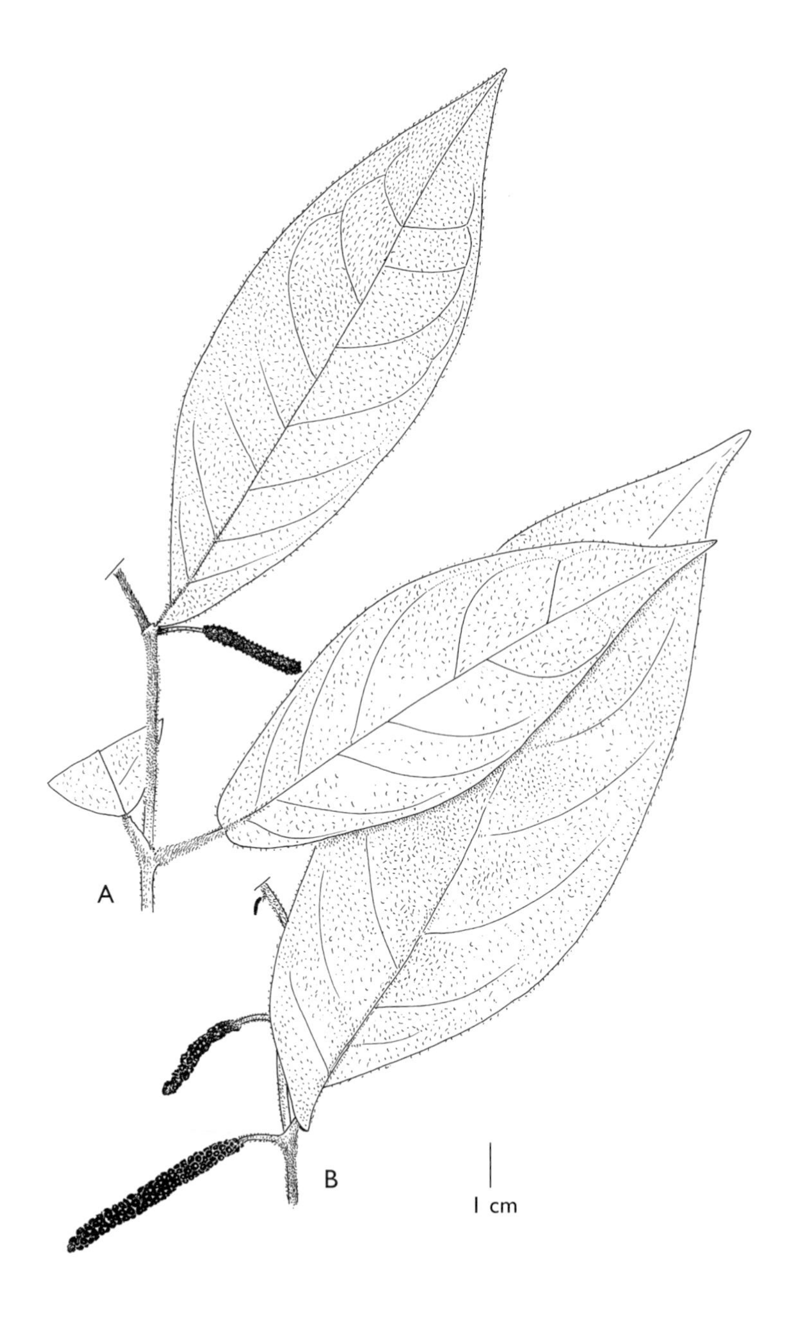

Fig. 31. *Piper brownsbergense* Yunck.: A, B, habit (A, de Granville *et al.* 4204; B, van Donselaar 1405).

secondary veins 4-7(-9) per side, originating from throughout primary vein, tertiary venation widely reticulate. Inflorescence pendent under leaves; peduncle 1-2 cm long, pilose, glabrescent; spike to 5.5 cm long; floral bracts triangular-subpeltate, marginally fringed. Infructescence pendent; fruits depressed-globose to obovoid, glabrous or papillate at apex, substylose, stigmas 3.

Distribution: The Guianas and Brazil (Amazonas); from sea level to 700 m elev.; 41 collections studied (GU: 5; SU: 9 ; FG: 27).

Selected specimens: Guyana: Upper Takutu - Upper Essequibo region, Rewa R., Clarke 3724 (U, US); Potaro-Siparuni region, W of Maikwak, Hahn 4223 (U, US). Suriname: Brokopondo Distr., Gansee, afterwards lake, van Donselaar 1405 (U); Lely Mts., Lindeman & Stoffers *et al.* 123 (U). French Guiana: Massif des Emerillons, de Granville *et al.* 3874 (CAY, P, U); Mt. Bellevue de l'Inini, de Granville *et al.* 7721 (CAY, NY, P, U).

Note: *Piper brownsbergense* and *P. hirtilimbum* are both pubescent with rather long hairs. *P. hirtilimbum* can be distinguished by the longer (to 2.5 mm) hairs on petioles and blades and the pubescent ovaries and fruits, whereas in *P. brownsbergense* the hairs on stem and petiole are at most 2 mm long and the ovaries and fruits are glabrous.

16. **Piper cernuum** Vell., Fl. Flum. 25. 1829 ('1825'); Icon 1: t. 58. 1831 ('1827'). Type: Brazil, near Rio de Janeiro, Pohl 28 (isotype K).

Shrub or treelet, 2-6 m tall. Stem densely pubescent. Petiole vaginate to apex, 4-9 cm long, densely pubescent; blade not scabrous, not glandular-dotted, broadly ovate to elliptic-ovate, 23-60 x14-28 cm, apex acute, base equal or almost equally attached to petiole, deeply unequally lobed, glabrous or sparsely pubescent above, veins glabrescent above, densely pubescent below; pinnately veined, secondary veins 5-8 per side, originating from lower 3/4 or more of primary vein, flat above and prominulous below, tertiary veins widely reticulate or sometimes transverse. Inflorescence pendent; peduncle 1-3(-5) cm long, densely pubescent; spike 20-60 cm long, yellow, not apiculate; rachis glabrous; floral bracts marginally fringed; stamens laterally dehiscent. Fruits obovoid, ca. 2-2.5 mm wide, puberulent at apex, stigmas 3, sessile, to 8 mm long.

Distribution: Colombia, Peru, Venezuela, Brazil, Guyana and French Guiana; in moist forest, swampy vegetation and stream sides,from sea level to 700 m elev.; 10 collections studied (GU: 3; FG: 5).

Selected specimens: Guyana: Potaro-Mazaruni region, Kaieteur Falls, Hahn 4178, 4654, 4728 (U, US). French Guiana: Saül, de Granville *et al.* B-5436 (CAY, U); foot of Mt. Galbao, de Granville *et al.* 8434 (CAY); sommet Tabulaire, Cremers *et al.* 6513 (CAY, U).

Notes: Steyermark (1984: 367-373) discerned several varieties and formas within *Piper cernuum*. With the small amount of Guianan specimens available it is impossible to distinguish different entities here. *Piper cernuum* and *P. obliquum* are very similar morphologically. Distinguishing characters are mentioned in the note to *P. obliquum*.

17. **Piper ciliomarginatum** Görts & Christenh., Blumea 50: 369. 2005. Type: Guyana, Kaieteur National Park, Henkel 2231 (holotype U, isotype US). – Fig. 32 A-F

Shrub to 1-3 m tall. Stem glabrous. Petiole of lower leaves 0.4-3 cm long, of upper leaves 0-0.3 cm long, glabrous, vaginate or winged to apex; blade not scabrous, densely glandular-dotted, elliptic-ovate to broadly ovate, 13-22 x 6-11 cm, margin ciliate, apex acute to acuminate, base equal or somewhat unequally attached to petiole difference 0-4 mm, acute or obtuse or rounded or subcordate, glabrous on both surfaces; pinnately veined, secondary veins 4-6 per side, originating from lower 1/2 of primary vein, flat above and below, tertiary veins reticulate. Inflorescence erect; peduncle 0.8-1.6 cm long, glabrous; spike 2-8 cm long, white, not apiculate; floral bracts densely marginally fringed. Infructescence pendent, 7-8 cm long, 3 cm thick, green; fruits trigonous, 0.4 mm wide, glabrous, stigmas 3, sessile.

Distribution: Venezuela, Guyana and Suriname; to 850 m elev.; 6 collections studied (GU: 4; SU: 1).

Selected specimens: Guyana:, Kaieteur National Park, Hahn 4669, Henkel 2166, 2185 (all U). Suriname: Wilhelmina Mts., Irwin *et al.* 54998 (K, U).

18. **Piper consanguineum** (Kunth) Steud., Nomencl. Bot. ed. 2. 2: 340. 1841. – *Steffensia consanguinea* Kunth, Linnaea 13: 623. 1840 ('1839'). Type: French Guiana, Poiteau s.n. (holotype G-DEL, isotype U). – Fig. 33 A-B

Artanthe leprieurii Miq., Syst. Piperac. 525. 1844. – *Piper leprieurii* (Miq.) Pulle, Enum. Vasc. Pl. Surinam 141. 1906. Type: French Guiana, Leprieur s.n. (holotype G-DEL, isotype U).

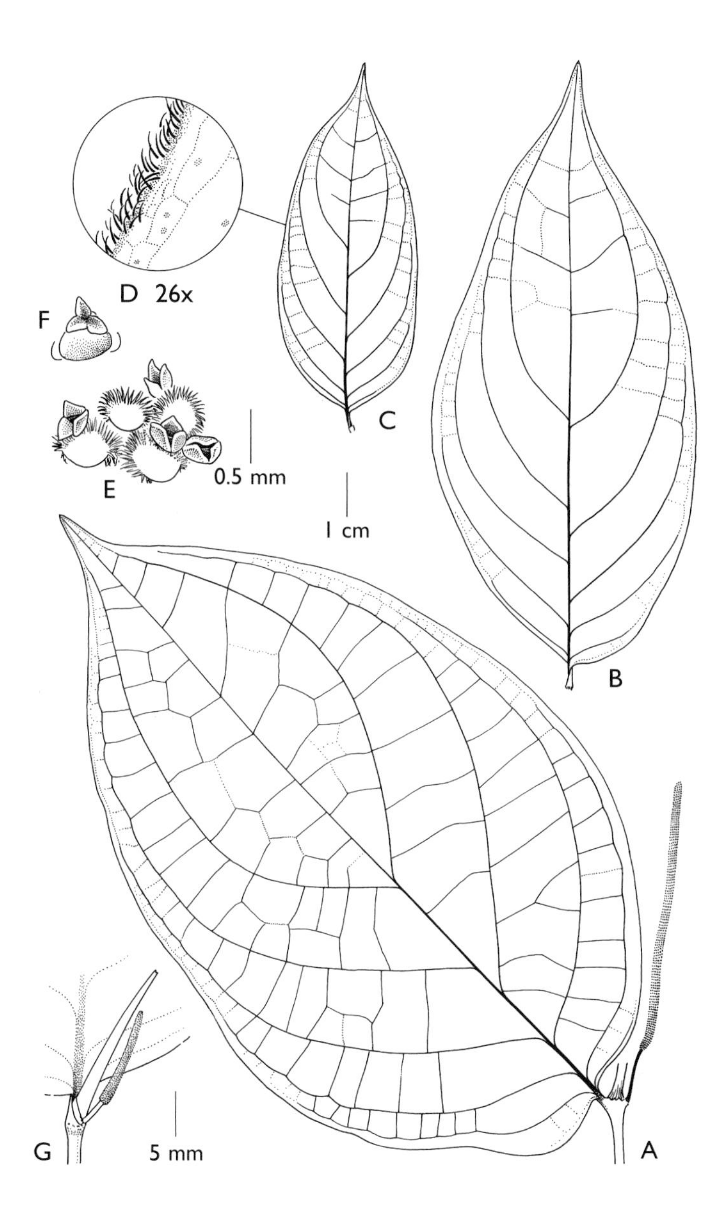

Fig. 32. *Piper ciliomarginatum* Görts & Christenhusz: A, apex of branch with leaf, withering prophyll and young spike; B-C, showing variation in leaf shape; D, detail of leaf margin; E, detail of spike showing floral bracts and young fruits with persistent stigma's; F, fruit with persistent stigma's; G, tip of young branch showing leaf base, prophyll and developing spike (A-B, G, Henkel 2231; C-D, F, Henkel 2166; E, Henkel 2185).

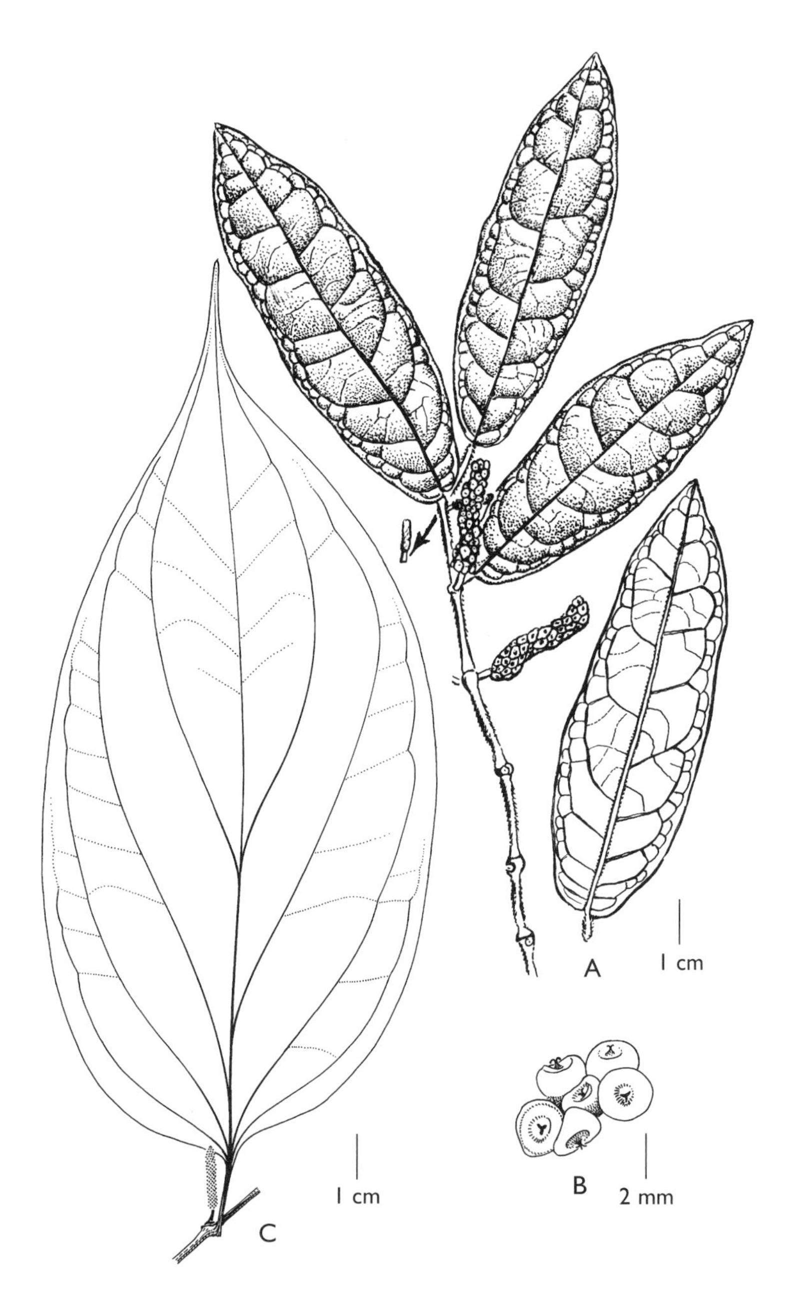

Fig. 33. *Piper consanguineum* (Kunth) Steud.: A, habit; B, fruits. *Piper crassinervium* Kunth: C, habit (A, drawing by W.H.A, Hekking; B, de Granville *et al.* 11757; C, de Granville *et al.* 7188).

Piper surinamense C. DC. in A. DC., Prodr. 16(1): 296. 1869. Type: not designated.

Herbaceous small shrub, 0.1-0.5 m tall, sometimes creeping, retrorsely crisp-pubescent. Petiole 0.2-1.5 cm long, crisp-pubescent, vaginate near base; blade dark green above often with a pale or white band along primary vein, pale green below, not conspicuously glandular-dotted, narrowly elliptic or lanceolate-oblong, 3.5-7(12) x 1-2(3.5) cm, apex obtuse-acute to acuminate, base equal or almost equally attached to petiole, rounded to subcordate, glabrous above, crisp-pubescent below especially on veins; pinnately veined, secondary veins 6-12 per side, originating from throughout primary vein, clearly anastomosing within margin. Inflorescence pendent; peduncle to 0.5 cm long, glabrescent; spike erect, 1.5-2.5 cm long, white, yellow to green, apiculate; floral bracts cucullate, glabrous; rachis pubescent. Infructescence 1-3 x 0.8-1 cm, green or brownish; fruits depressed globose, to 3 mm in diam., papillose, green, becoming exserted, stigmas 3, recurved on a very short style.

Distribution: SE Venezuela, the Guianas and Brazil (Amazonas, Roraima, Rondônia, Pará); understorey of primary forest or swampy forest, or open forest on laterite; up to 650 m elev.; ca. 130 collections studied (GU: 10; SU: 24; FG: 96).

Selected specimens: Guyana: Rupununi Distr., Kuyuwini Landing, Jansen-Jacobs *et al.* 3102 (P, U); Kuyuwini R., Clarke 4538 (U, US). Suriname: Wilhelmina Mts., above confluence Lucie R., Irwin 55779 (NY, U); Suriname R., Jodensavanne-Mapane Cr. area, Schulz LBB 8291 (BBS, U). French Guiana: Station des Nouragues, Cremers *et al.* 10935 (CAY, NY, P, U, US); Approuague R., Ineri Mt., Cremers *et al.* 15352 (CAY, P, U, US).

Vernacular names: Suriname: snekibita (Sranan; LBB 8291), nowtu (Ndjuka; Daniel 50).

Use: According to Schulz, Suriname Amerindians use leaves extracted in spirit after snake bite (see also the vernacular name, meaning snake bite).

Notes: This species is easily recognizable by the many anastomosing secondary veins and the often variegate leaf pattern.
Several collections (e.g. Suriname, Wessels Boer 1533 and Venezuela, Morillo 9024) are quite similar to *Piper consanguineum* but differ in the size of the leaves up to 16 x 5 cm and glabrous somewhat abortive mature spikes. It may turn out that they belong to another taxon which can only be decided when more material is available.

19. **Piper coruscans** Kunth in Humb., Bonpl. & Kunth, Nov. Gen. Sp. ed. qu. 1: 53. 1816. Type: Colombia, Antioquia, Magdalena R., between S. Bartholomeo and Garapatas, Humboldt 1629 (holotype P, isotype B-W 672, none seen). – Fig. 34 A

Piper pseudochurumayu (Kunth) C. DC. var. *membranaceum* C. DC. in A. DC., Prodr. 16(1): 288. 1869. – *Piper coruscans* Kunth var. *membranaceum* (C. DC.) Steyerm., Fl. Venez. 2(2): 383. 1984. Type: Venezuela, Amazonas, Spruce 3433 (holotype G-DC, isotype K, none seen).
Piper orenocanum C. DC. var. *koriaboense* Trel. & Yunck., Piperac. N. South Amer. 285. 1950. Type: Guyana, NW Distr., near Koriabo, Archer 2505 (not 2506 as cited in Trel. & Yunck.) (holotype US, not seen, isotype K).

Shrub or subshrub to 1-3 m tall. Stem glabrescent. Stipules about as long as petiole. Petiole 2-3.2(-6.5) cm long, glabrous or sparsely pubescent, vaginate to apex; blade firmly membranous, often drying yellowish or greyish green, not scabrous, not conspicuously glandular-dotted below, broadly ovate to ovate, or even suborbiculate, 11-24 x 6-13 cm, apex long-acuminate, base equal or almost equally attached to petiole, truncate, rounded or cordulate, glabrous above or primary vein somewhat appressed-pubescent, veins appressed-pubescent below; pinnately veined, secondary veins 4-7 per side, originating from lower 1/2 or lower 2/3 of primary vein, ascending at an angle of 40-45°, plane to impressed above, prominent below, tertiary veins obsoletely transverse. Inflorescence erect; peduncle 0.5-1(-2.5) cm long, glabrous; spike 8-11 cm long, white, apiculate; floral bracts rounded to trigonous, densely marginally fringed. Infructescence erect, green; fruits obovoid or somewhat angular, depressed at apex, 0.8-1 mm wide, glabrous, green, stigmas 3, sessile.

Distribution: Colombia, Venezuela, Guyana, French Guiana and W Brazil, Ecuador, Bolivia; riverine flood plain forest on white and brown sands, from mud in wet forest on granitic rocks, red clay or deep mud along river, also from swamp forest, from sea level to 650 m elev.; 15 collections studied (GU: 10; FG: 1).

Selected specimens: Guyana: Barima-Waini region, Barima R. head water in small rapid, Pipoly 8277 (BBS, NY, U, US), Tenapu Cr., E of Arakaka, Pipoly 8102 (NY, U, US); Cuyuni-Mazaruni region, along Koatse R., Pipoly 10784 (NY, U, US). French Guiana: Saül, Görts *et al.* 62 (CAY, NY, U).

Note: For distinguishing characters between *Piper coruscans, P. glabrescens* and *P. perstipulare*, see note to the latter species.

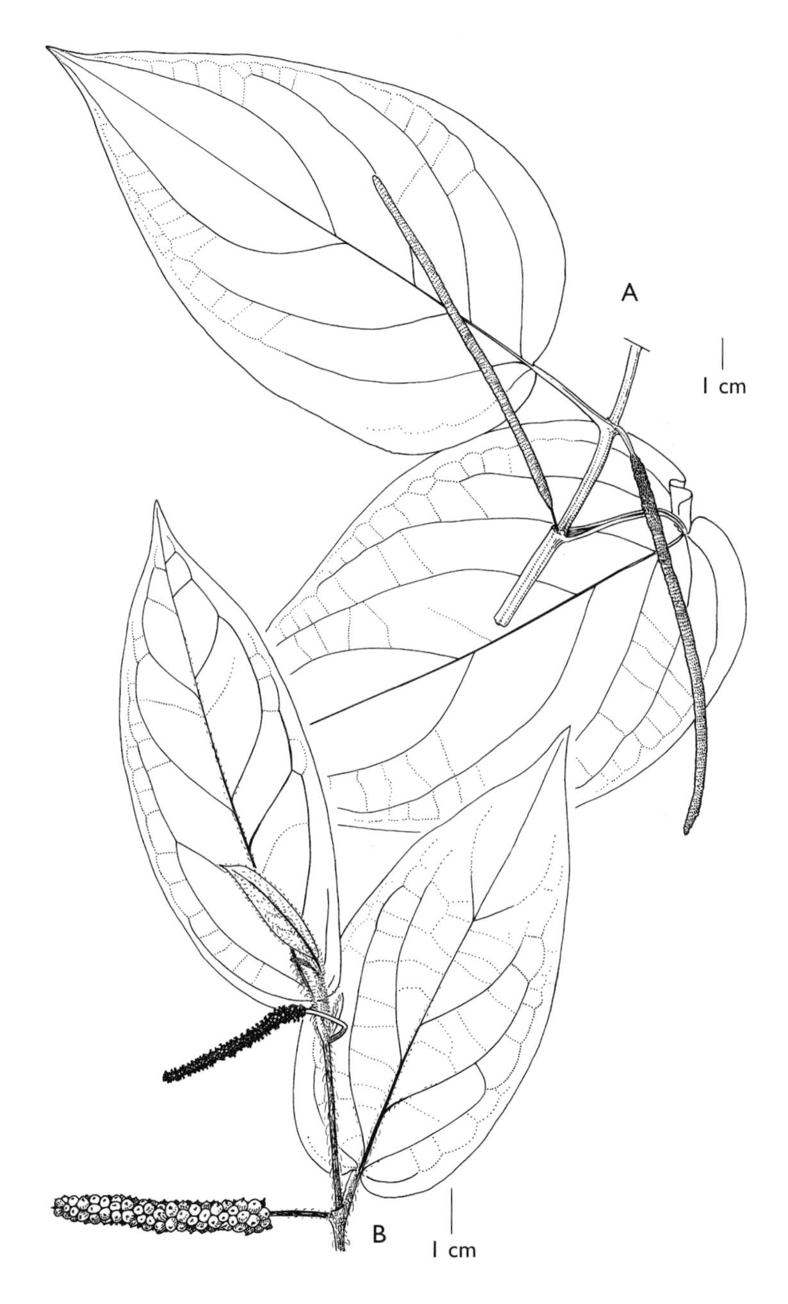

Fig. 34. *Piper coruscans* Kunth: A, habit. *Piper cyrtopodum* (Miq.) C. DC.: B, habit (A, Pipoly 8277; B, Maas *et al.* 3507).

20. **Piper crassinervium** Kunth in Humb., Bonpl. & Kunth, Nov. Gen. Sp. ed. qu. 1: 48. 1816. Type: Colombia. Magdalena R., near Honda, Humboldt 655 (holotype P, isotype B-W 655, none seen).
– Fig. 33 C

Shrub or treelet, 2-5 m tall. Petiole ca. 1.5 cm long, glabrous or pubescent, vaginate to middle or to apex; blade not scabrous, not glandular-dotted, ovate, 13-20 x 5-13 cm, apex acuminate, base equal or almost equally attached to petiole, truncate or cuneate, but acute to petiole, glabrous above, veins more or less pubescent below; pinnately veined, secondary veins 4-5 per side, originating from lower 1/2 of primary vein, prominent below. Inflorescence: peduncle stout, 0.5-1.5 cm long, pubescent; spike 12-15 x 0.5-0.7 cm; floral bracts marginally fringed. Fruits rounded, glabrous, with persistent style, stigmas 3, recurved.

Distribution: Costa Rica, Colombia, Ecuador, Peru, the Guianas and Brazil; 5 collections studied (GU: 1; SU: 1; FG: 2).

Specimens examined: Guyana: Upper Takutu-Upper Essequibo region, Kamoa Mts., Clarke 3106 (U. US). Suriname: Wilhelmina Mts., lower slopes of Julianatop, Irwin *et al.* 54816 (U). French Guiana: Upper Camopi R., Mt. Belvédère, de Granville *et al.* 7188 (CAY, NY, U, US); Mt. Atachi Bacca, de Granville *et al.* 10899 (P, U).

21. **Piper cyrtopodum** (Miq.) C. DC. in A. DC., Prodr. 16(1): 337. 1869, as 'cyrtopodon'. – *Peltobryon cyrtopodum* Miq. in Mart., Fl. Bras. 4(1): 219. 1853. Type: Brazil, Pará, Obidos, Spruce s.n. (holotype K, isotype BM). – Fig. 34 B

Piper kappleri C. DC., J. Bot. 4: 213. 1866. – *Piper marowinianum* C. DC. in A. DC., Prodr. 16(1): 337. 1869, nom. illeg. Type: Suriname, Marowijne R., Kappler 1885 (on p. 213 as '1855') (holotype G not seen, isotypes P, U), syn. nov.
Piper subciliatum C. DC., Annuaire Conserv. Jard. Bot. Genève 2: 263. 1898. Type: French Guiana, Leprieur 148 (holotype G-DEL, isotype P), syn. nov.

Subshrub or herb, 1-3(-4) m tall. Stem pilose with hairs to 2 mm long. Petiole 0.2-1 cm long, pilose, glabrescent, vaginate to apex, sometimes reddish tinged; blade somewhat scabrous, not conspicuously glandular-dotted, elliptic-ovate, elliptic-oblong or lanceolate, 9-20 x 4-6(-10) cm, margin hardly ciliate, apex acuminate or acute, base equal or almost equally attached to petiole, subcordate to rounded, somewhat pilose above, more so below; pinnately veined, secondary veins (4-)6-8 per side, originating from throughout primary vein, plane above, prominent below. Inflorescence pendent; peduncle 0.9-1.6 cm long, pilose, often red; spike 2.5-5 cm long, not apiculate; floral bracts cucullate and pilose on inner side and at base. Fruits ovoid, ca. 2 mm wide, glandular, glabrous, stigmas 3, may be on a short style or fruits apiculate.

Distribution: The Guianas and Brazil (Amazonas and Pará); virgin forest, mossy cloud forest, forest on granite and from open places, up to 700 m elev.; 29 collections studied (GU; 6; SU: 7; FG: 16).

Selected specimens: Guyana: Thompson's farm, Maas & Westra 3507 (BBS, BRG, NY, U); Mabura, W Pibiri compartment, Ek 1255 (BRG, U). Suriname: Along Toso Cr., near Pasjansi, Florschütz 381 (U); Nassau Mts., Lanjouw & Lindeman 2110 (U). French Guiana: Sinnamary R., Petit Saut, Prévost 1423 (CAY, NY, U); Mt. Bellevue de l'Inini, de Granville *et al.* 8036 (CAY, P, U).

Notes: *Piper marowinianum* is based on among others Kappler 1885, it thus is a nomenclatural synonym of *P. kappleri*.
Piper subciliatum is only known from its type collection. This specimen fits well within *P. cyrtopodum*.

22. **Piper demeraranum** (Miq.) C. DC. in A. DC., Prodr. 16(1): 298. 1869. – *Artanthe demerarana* Miq., London J. Bot. 4: 464. 1845. Type: Suriname, Hostmann 312 (lectotype and isolectotypes K) (designated by Trel. & Yunck. 1950: 380). – Fig. 35 A-D

Piper lenormandianum C. DC. in A. DC., Prodr. 16(1): 299. 1869. Type: French Guiana, Acarouany, Sagot 540 (holotype K, isotypes B?, BM, P), syn. nov.
Piper rubescens C. DC. in A. DC., Prodr. 16(1): 300. 1869. – *P. demeraranum* (Miq.) C. DC. var. *rubescens* (C. DC.) Trel. & Yunck., Piperac. N. South Amer. 380. 1950. Type: French Guiana, Acarouany, Sagot 844 (holotype B, not seen, isotype K).

Subshrub or shrub, 1-2.5 m tall. Stem crisp-pubescent or sometimes appressed-pubescent. Petiole 0.3-0.5(-1) cm long, crisp-pubescent, vaginate to blade; blade membranous or subcoriaceous, not scabrous, not glandular-dotted, dark green above, more pale below, elliptic to oblanceolate, 11-26(-35) x 4-9 cm, apex acuminate, base unequally attached to petiole difference 0.1-0.4 cm, unequally cordate, with slightly larger, auriculate lobe mostly covering petiole, glabrous above, crisp-pubescent below; pinnately veined, secondary veins 8-11(-14) per side, originating from throughout primary vein, flat to impressed above, prominent below, tertiary veins widely reticulate. Inflorescence erect;

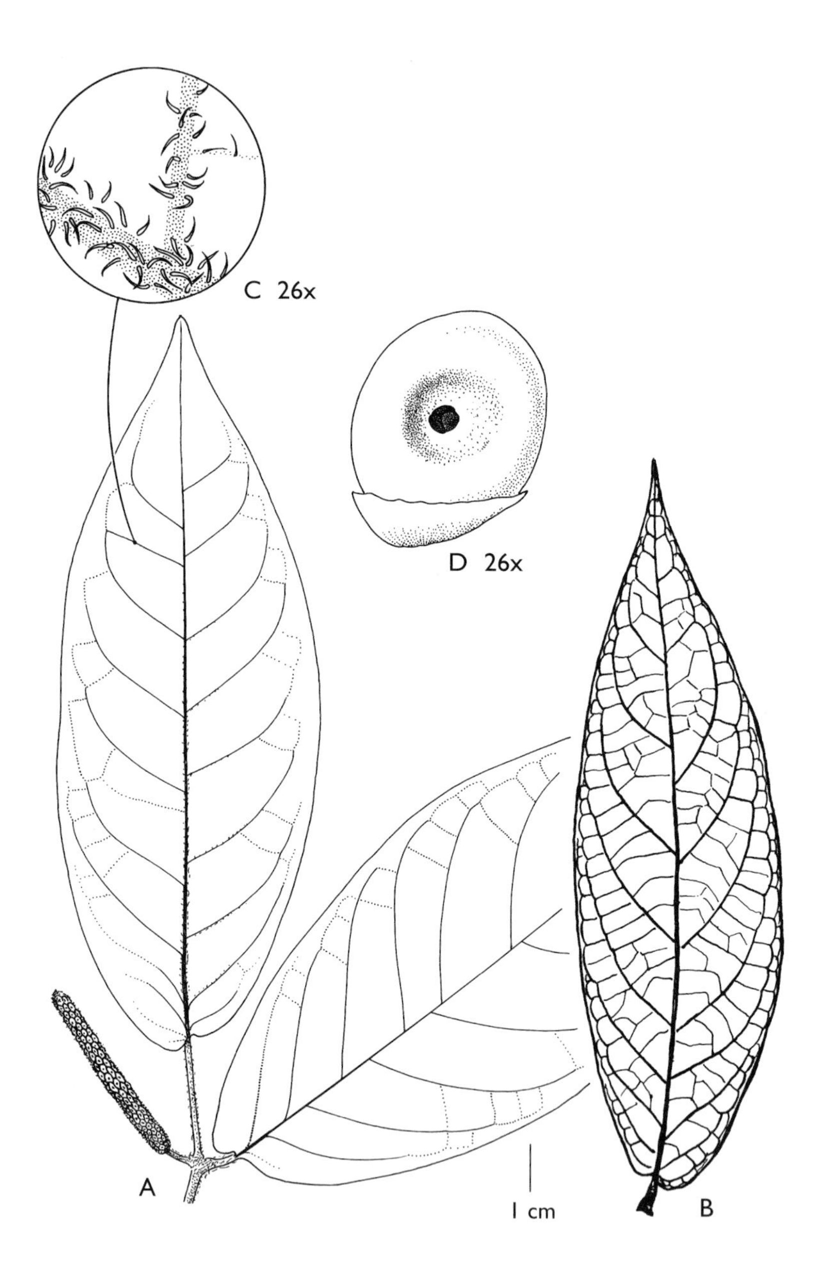

Fig. 35. *Piper demeraranum* (Miq.) C. DC.: A, habit; B, leaf; C, detail of leaf upper surface indument; D, fruit with subtending floral bract (A, C-D, Görts-van Rijn *et al.* 102; B, Lindeman 5082). B, drawing by W.H.A. Hekking.

peduncle to 1 cm long, pubescent, green; spike 3.5-4.5 cm long, white, yellow or green, apiculate; rachis glabrous; floral bracts cucullate, glabrous or dorsally puberulent. Infructescence to 6 cm long, 10 mm wide, green; fruits globose, 2-2.5 mm wide, puberulent at apex, stigmas 3, sessile.

Distribution: Colombia, Venezuela, Trinidad, the Guianas and Brazil (Amazonas); in dense forest, montane forest, growing on brown sand with granite boulders or on sandstone, from 20-720 m elev.; 52 collections studied (GU: 25; SU: 14; FG: 16).

Selected specimens: Guyana: NW Kanuku Mts., A.C. Smith 3612 (K, NY, U); Mt. Makarapan, Maas *et al.* 7535 (BRG, U). Suriname: Kabalebo dam area, Lindeman & Görts-van Rijn *et al.* 244 (BBS, NY, U); Perica R., Commewijne Distr., Lindeman 5105 (BBS, NY, U). French Guiana: Saül area, Görts-van Rijn *et al.* 102 (CAY, NY, U); Trinité Mts., de Granville *et al.* 6467 (BR, CAY, G, MG, P, U).

Vernacular name: Guyana: warakabakoro (Fanshawe 792). This name is not unique for the taxon.

Notes: According to Trelease and Yuncker (1950: 380), specimens in NY with the type number, Hostmann 312, belong to *Piper hispidum* var. *trachydermum.*
Trelease and Yuncker (1950: 378) placed *Piper lenormandianum* C. DC. in synonymy of *P. kappleri*. While studying the Kew type specimen of the latter I had to conclude that this specimen is clearly different and belongs to *P. demeraranum* instead of to *P. cyrtopodum.*
The numbering of the Sagot collections poses a problem here. Sagot 844 at K is certainly *P. demeraranum.* Specimens with the same number 844 in NY and P are correctly assigned to *P. wachenheimii.* The two species are clearly separable by the shape of the leaves. We have to conclude that 844 is not an exclusive number for the species mentioned.

23. **Piper dilatatum** Rich., Actes Soc. Hist. Nat. Paris 1: 105. 1792.
Type: not designated. – Fig. 36 A-B

Shrub or subshrub to 4 m tall. Stem crisp-pubescent or scabrous. Prophyll densely pubescent. Petiole 0.5-1.5 cm long, pubescent sometimes densely so, vaginate to apex, occasionally reddish tinged; blade hardly to densely scabrous above, not glandular-dotted, rhombic, elliptic to subobovate, 15-20 x 7-9 cm, apex acuminate, base unequally attached to petiole difference 0.3-0.4 cm, unequally cordate, rounded or obtusish, upper surface glabrous to pubescent, veins pubescent, lower

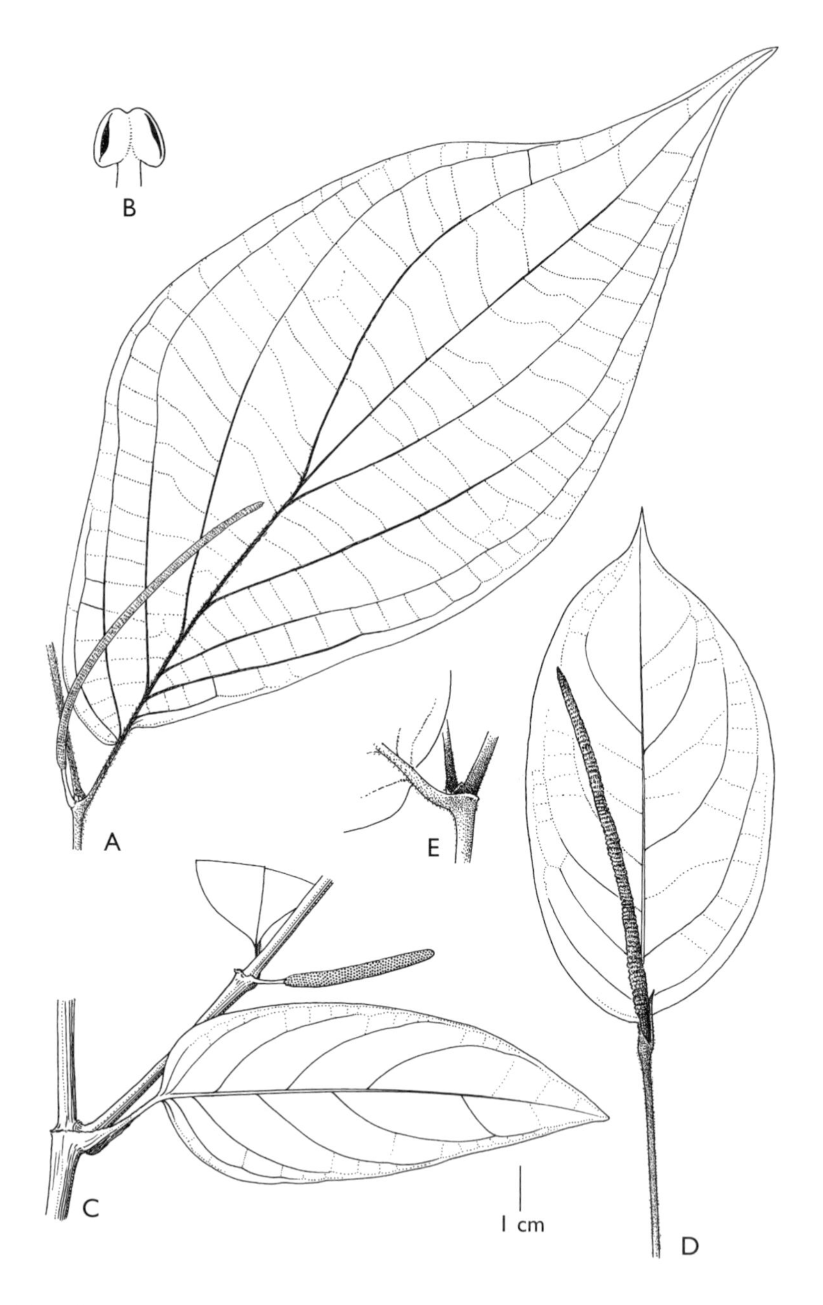

Fig. 36. *Piper dilatatum* Rich.: A, habit; B, stamen. *Piper divaricatum* G. Mey.: C, habit. *Piper duckei* C. DC.: D, habit; E, detail of leaf attachment (A-B, de Granville *et al.* 66; C, Gillespie 1008; D-E, Clarke 4385).

surface sparsely to densely pubescent; pinnately veined, secondary veins 5-6 per side, originating from lower 1/2 of primary vein, tertiary veins transverse and parallel reticulate. Inflorescence erect; peduncle 0.5-1 cm long, puberulent to glabrous; spike 7-8 cm long, white to yellow, not apiculate; floral bracts 0.7-0.8 mm in diam., densely marginally fringed; stamen connective narrow, anthers dehiscing laterally. Fruits trigonous, glabrous, or with some hairs at top, stigmas 3, sessile.

Distribution: West Indies and northeastern S America; 43 collections studied (GU: 20; SU: 6; FG: 17).

Selected specimens: Guyana: Kanuku Mts., Turuk wau, Cook 99 (K, NY, U); Rupununi Distr., Rupununi R., Kuyuwini Landing, Jansen-Jacobs *et al.* 3269 (BBS, P, U). Suriname: Jenny, Coronie, LBB 15152 (BBS, U); Tawajariweg, LBB 16412 (BBS). French Guiana: Saül area, Görts-van Rijn *et al.* 66 (CAY, NY, U, US), Barthelemy 151 (CAY, NY, U).

Vernacular name: Suriname: aneisie wiwiri (Reijenga 657). This name is not unique for the taxon.

Notes: *Piper dilatatum* and *P. hispidum* have superficial resemblance; they differ, however, in the shape and the scabrity of the leaves. In *P. dilatatum* the leaves are mostly rhombic and not strongly scabrous, whereas in *P. hispidum* they are elliptic or ovate and strongly scabrous. In *P. dilatatum*, the anthers open laterally and the connective between the thecae is not broadened. In *P. hispidum*, the connective is broadened and the anthers open apically. The prophyll in *P. dilatatum* is densely pubescent, while in *P. hispidum* it is glabrous with pubescent midvein. Burger (1971: 128) stated that the two taxa differ significantly in anther and fruit characters.
Piper pseudofuligineum C. DC. is closely related to *P. dilatatum*. It is a Central American taxon, maybe a somewhat more pubescent form adapted to drier habitats (according to Burger).
Piper taboganum C. DC. is considered conspecific with *P. pseudofuligineum*. The A.C. Smith collections 2416, 3451 and 3079 from S Guyana identified as *Piper taboganum* seem better assigned to *P. dilatatum* and fit well into the distribution of the latter.
Sandwith 1544 from Kyk-overal on the Mazaruni R., Guyana was identified as *Piper grahamii* Trel. It probably belongs to *P. dilatatum*. A final conclusion on *P. grahamii* as a synonym of *P. dilatatum* can only be drawn after study of Graham 277, the type specimen that was collected in the same locality.
Yuncker (1957: 233) cited Splitgerber 225 (L) as the only Suriname report of *P. auritum* Kunth; this collection, however, has to be included in *P. dilatatum*. *Piper auritum* is known from Mexico to Colombia.

24. **Piper divaricatum** G. Mey., Prim. Fl. Esseq. 15. 1818. Type: Suriname, Weigelt s.n. (holotype W, not seen, photo SP).
– Fig. 36 C

Piper romboutsii Yunck. in Trel. & Yunck., Piperac. N. South Amer. 395. 1950. Type: Suriname, Corantijne R., Rombouts 192 (holotype U), syn. nov.

Shrub, subshrub or treelet, sometimes scandent, to 3 m tall. Stem glabrous. Petiole 1-2.2 cm long, vaginate or grooved to apex; blade coriaceous, drying shiny or membranous, dark glandular-dotted on both sides, rhombic, somewhat asymmetric, 10-20 x 3-8 cm, apex acute to acuminate, base almost equal, obtuse or acute, glabrous; pinnately veined, secondary veins 4-6 per side, originating from lower 3/4 of primary vein, at a wide angle and abruptly curving upward, not anastomosing, shallowly or deeply impressed above, prominulous to prominent below, tertiary veins reticulate, slightly prominulous or inconspicuous. Inflorescence usually pendant; peduncle 0.5-0.8(-1.5) cm long, glabrous; spike 3-6 cm long, whitish, pale yellow or green, apiculate; floral bracts marginally fringed. Infructescence 6 x 1 cm (when dried); fruits obovoid, 1-2 mm thick, glabrous, becoming exserted, stigmas 3, sessile.

Distribution: W Venezuela, the Guianas, Brazil (Amazonas, Pará and Amapá) and Bolivia; in riverine forest, swamp forest, occasionally reported from regenerating forest; over 80 collections studied (GU: 20; SU: 42; FG: 15).

Selected specimens: Guyana: Molson Cr., Corantijn R., R. Persaud 257 (BRG, U); Demerara R., Jenman 4206 (BM, BRG, U). Suriname: near Paramaribo, between Houttuin and Lelydorp, Hekking 912 (BBS, NY, U); Lucie R., at confluence with Oost R., Irwin *et al.* 55470 (K, NY, U). French Guiana: Maroni R., Prévost 1722 (BM, CAY, NY, P, U); Upper Marouini R. Basin, de Granville *et al.* 10030 (CAY, NY, P, U, US).

Vernacular names: Suriname: mböma (u)wi (Sauvain 498, 616) or boma wi (Ndjuka; Sauvain 498); man-anesiwiwiri (Sranan, this name is not unique for the taxon).

Notes: In a personal communication, Callejas informed me that in the Western range of the species the fruits can be densely pubescent.
Trelease & Yuncker (1950: 227) mention in synonymy of *P. divaricatum*: "*? Piper praemorsum* Rottb. ex Vahl". This was validly published in Eclog. Amer. 1: 4. 1797. No type is known for *P. praemorsum*; according to Rottbøll, it was collected in Suriname by Rolander. If this species really would appear to be a synonym of *P. divaricatum*, *P. praemorsum* would have to replace *P. divaricatum*.

25. **Piper duckei** C. DC., Notizbl. Bot. Gart. Berlin-Dahlem 7(62): 446. 1917, as 'duckii'. Type: Brazil, Pará, Murutucú, Ducke & J. Huber MG 3400) (holotype G, isotype MG). – Fig. 36 D-E

Shrub to 1.5 m tall. Stem densely hirsute. Petiole 0.5-1.2 cm long, hirsute, broadly vaginate or winged to apex; blade drying glossy above, not scabrous, not conspicuously glandular-dotted, slightly asymmetrical, broadly elliptic, 13-21 x 7.5-10 cm, apex (long) acuminate, base equal or unequally attached to petiole difference 0-0.5 cm, rounded or cordulate, one side may be subacute, glabrous above, hirsute below, veins densely so; pinnately veined, secondary veins 4-8 per side, originating from lower 3/4 or 2/3 of primary vein, impressed above, tertiary veins reticulate, prominulous. Inflorescence erect; peduncle 0.4-1.5 cm long, crisp-pubescent; spike to 16 cm long, pink, apiculate; floral bracts densely marginally fringed. Fruits obpyramidate-trigonous, apex depressed, glandular, pubescent, stigmas 3, linear, sessile.

Distribution: Brazil (Amazonas and Pará) and Guyana; dense forest on brown sand, 240 m elev.; 2 collections studied (GU: 1).

Specimens examined: Guyana: Upper Takutu - Upper Essequibo region, Kuyuwini R. trail to Kassikaityu R., Clarke 4385 (U, US).

26. **Piper dumosum** Rudge, Pl. Guian. Rar. 13, t. 14. 1805. Type: French Guiana, J. Martin s.n. (holotype BM). – Fig. 37 A

Artanthe adenophora Miq., Syst. Piperac. 514. 1844. – *Piper adenophorum* (Miq.) C. DC. in A. DC., Prodr. 16(1): 274. 1869. Syntypes: French Guiana, Poiteau s.n. (G-DEL not seen, U) and Leprieur 153 (G-DEL not seen).

Shrub or treelet to 3 m tall. Stem glabrous, densely dark glandular-dotted. Stipules large 3 x 1.5 cm, early caducous. Petiole 0.5-2(to 6 in fertile twigs) cm long, dark-villous (hairs at least 1 mm long), vaginate to middle, green to reddish; blade may be somewhat bullate *in vivo,* glandular-dotted, lance-elliptic to ovate to broadly ovate 18-22(-27) x 5.5-7.5(-17) cm, apex short-acuminate, base equal or almost equally attached to petiole, acute, truncate to subcordate, sparsely villous above, more densely so below; pinnately veined, secondary veins 5-6 per side, originating from lower 3/4 of primary vein. Inflorescence pendent; peduncle 1-1.2 cm long, slender, glabrous, green to reddish; spike 5-8 cm long, white, yellowish to green, not apiculate; floral bracts densely marginally fringed. Fruits obovoid to trigonous, glabrous, depressed at apex when dried, stigmas 3, erect.

Fig. 37. *Piper dumosum* Rudge: A, habit. *Piper eucalyptifolium* Rudge: B, habit (A, de Granville *et al.* 8960; B, de Granville *et al.* 9051).

Distribution: The Guianas and adjacent Brazil; mainly from dense forest, or swampy places, up to 700 m elev.; over 70 collections studied (GU: 1; SU: 3; FG: 62).

Selected specimens: Guyana: Barima-Waini region, Pipoly 8023 (BRG, NY, U, US). Suriname: Litani R., Rombouts 904 (NY, U); Tumac Humac Mts., Telouakem, Acevedo 5837 (U). French Guiana: Mt. Chauve, Cremers *et al.* 15127 (B, CAY, NY, MO, MY, P, U, US); Saül area, de Granville *et al.* 8516 (CAY, NY, P, U, US).

Vernacular names: French Guiana: yakamilepia (Wayampi; Grenand 733, Jacquemin 2361, de Granville *et al.* B-5209); yalitakuã (Wayampi; Jacquemin 2024).

Note: Easily recognizable by the glabrous stem and sparsely villous leaves and peduncles; very densely dark-glandular-dotted all over.

27. **Piper eucalyptifolium** Rudge, Pl. Guian. Rar. 10, t. 6. 1805. Type: French Guiana, J. Martin s.n. (holotype BM). – Fig. 37 B

Artanthe rhynchostachya Miq., Syst. Piperac. 526. 1844. – *Piper rhynchostachyum* (Miq.) C. DC. in A. DC., Prodr. 16(1): 326. 1869. Type: French Guiana, Poiteau s.n. (holotype G-DEL, not seen, isotypes P, U).

Shrub or treelet to 2 m tall, glabrous except for subnodal lines of hairs on stem. Petiole 0.3-0.7 cm long, ciliate, vaginate to apex; blade not scabrous, not conspicuously glandular-dotted, dark green above, pale below, narrowly lanceolate, 6-16 x 1.5-3 cm, apex long-acuminate, base almost equal, acute, glabrous; pinnately veined, secondary veins 8-15 per side, originating from throughout primary vein, plane above, slightly prominulous below, tertiary veins obsolete. Inflorescence erect; peduncle ca. 0.5 cm long, glabrescent; spike 0.5-3 cm long, apiculate; floral bracts cucullate, inflexed, glabrous. Fruits depressed globose, glabrous, stigmas 3(-4), sessile.

Distribution: French Guiana and Brazil; in dense forest understorey, up to 639 m elev.; 24 collections studied (FG: 24).

Selected specimens: French Guiana: Inini Mt., Inini R. basin, Feuillet 3734 (CAY, U, US), de Granville *et al.* 7948 (CAY, NY, P, U), de Granville *et al.* 7492 (B, BR, CAY, G, K, MG, MO, P, U, US); Saül area, de Granville *et al.* 8416 (CAY, NY, P, U, US); Kaw Mts., Weitzmann 297 (CAY, U, US).

Note: For characters distinguishing this species from *Piper anonifolium* and *P. angustifolium* see notes under these species.

28. **Piper fanshawei** Yunck., Mem. New York Bot. Gard. 9: 321. 1957. Type: Guyana, Pakaraima Mts., Membaru-Kurupung Trail, Maguire & Fanshawe 32356 (holotype NY, isotype NY). – Fig. 38 A

Small shrub to 2 m tall. Stem glabrous. Petiole 1-2 cm long, glabrous, vaginate below middle; blade not scabrous, not glandular- but finely pellucid-dotted, elliptic(-oblong), 10-14 x 3-4.5 cm; margin not ciliate, apex acuminate, base equal or unequally attached to petiole difference 0.2 cm, acute, glabrous; pinnately veined, secondary veins 5-6 per side, originating from throughout primary vein, plane or inconspicuous above, slightly prominulous below, tertiary veins inconspicuous. Inflorescence erect; peduncle filiform, 1-1.5 cm long; spike 2 cm long, apiculate; floral bracts dorsally fringed, crowded. Fruits obovoid, puberulent at depressed apex, stigmas sessile.

Distribution: Only known from Guyana from 600-1500 m elev.; ca. 9 collections studied (GU: 9).

Selected specimens: Guyana: Cuyuni-Mazaruni region, Paruima, Clarke 5362 (U, US); Pakaraima Mts., Kurupung-Membaru trail, Hoffman 2158 (U), Mt. Ayanganna, Hoffman 3246 (U, US); Potaro-Siparuni region, Pakaraima Mts., Mt. Wokomong, Henkel 4420 (U, US).

Notes: Glabrous plants with pubescent fruits mark the species. It is difficult to identify collections without mature fruits.
Piper fanshawei can easily be mistaken for *P. cuyunianum* Steyerm. (1984: 391) from SW Venezuela, but differs in having pubescent fruits whereas *P. cuyunianum* has glabrous fruits.
Piper fanshawei differs from *P. bolivaranum* in the wider leaves: in *P. fanshawei* 3-4.5 cm wide, in *P. bolivaranum* 0.7-1.5 cm; *P. bolivaranum* also has glabrous - not pubescent - fruits.

29. **Piper flexuosum** Rudge, Pl. Guian. Rar. 13, t. 13. 1805) – *Arthanthe flexicaulis* Miq., Syst. Piperac. 533. 1844, nom. illeg. – *Piper flexicaule* C. DC. in A. DC., Prodr. 16(1): 308. 1869, nom. illeg. Type: French Guiana, J. Martin s.n. (holotype BM). – Fig. 38 B

Shrub to 2? m tall. Stem densely crisp- pubescent. Petiole 0.5 cm long, crisp-pubescent, vaginate to apex; blade not scabrous, glandular-dotted, elliptic, 8-13 x 3.5-5.5 cm, apex acute, base almost equally attached to

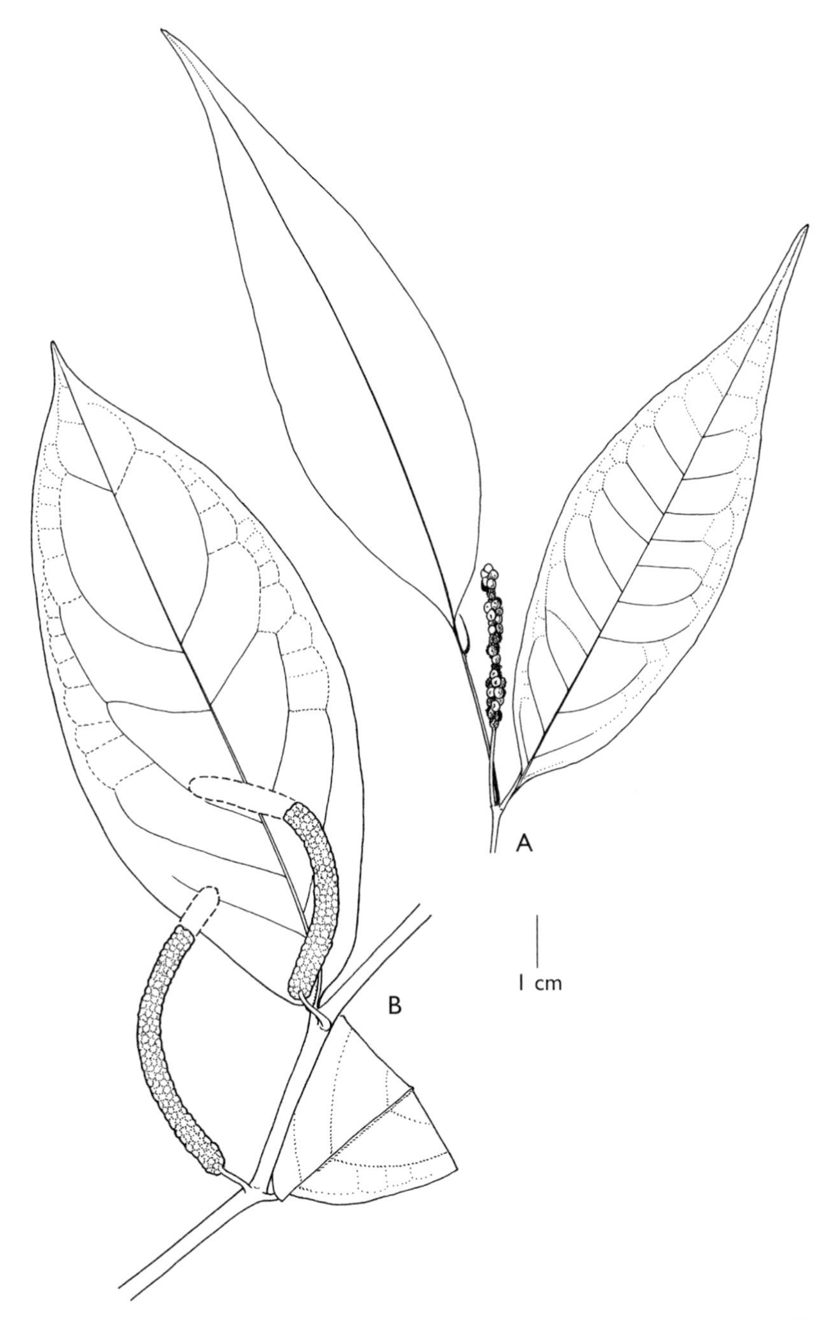

Fig. 38. *Piper fanshawei* Yunck.: A, habit. *Piper flexuosum* Rudge: B, habit (A, Henkel 4420; B, Martin s.n.).

petiole, unequally obtuse or acute, glabrous except for veins crisp-pubescent below; pinnately veined, secondary veins 5-7 per side, originating from throughout primary vein, prominent below, tertiary veins inconspicuously reticulate. Inflorescence erect or slightly curved; peduncle 0.5 cm long, crisp-pubescent; spike 4-6 cm long, 0.4-0.5 cm thick, apiculate; floral bracts densely marginally fringed. Fruits depressed globose or ovoid, glabrous, stigmas sessile.

Distribution: Guyana and French Guiana (GU: 3; FG: 2).

Specimens examined: Guyana: Bonasica, Kortright 8860 (K); Mazaruni Station, Forest Dept. 2588 (K); Demerara, Parker s.n. (K). French Guiana: Isle Bagatelle, Sinnamary R., Hoff 6372 (U).

Notes: "Demerara", Parker s.n. in Herbarium Hookerianum (Kew) is not the type of *Piper flexuosum*.
A later homonym exists: *P. flexuosum* J. Jacq., Eclog. Pl. Rar. 1: 139, t. 93. 1816. Miquel (1844) created much confusion when transferring both species to *Artanthe*. For Jacquin's species he made the combination *A. flexuosa* Miq., nom illeg. (p. 454), whereas for the oldest species he chose a new epithet: *A. flexicaulis* Miq., nom. illeg. (p. 533).

30. **Piper fuligineum** (Kunth) Steud., Nomencl. Bot. ed. 2. 2: 341. 1841. – *Steffensia fuliginea* Kunth, Linnaea 13: 655. 1840 ('1839'). Type: S Brazil, Sellow s.n. (holotype B not seen, isotype U). – Fig. 39 A-B

Artanthe amplectens Miq., Linnaea 20: 154. 1847. – *Piper amplectens* (Miq.) C. DC. in A. DC., Prodr. 16(1): 293. 1869. Type: Brazil, without precise locality, Pohl 1366 (holotype W, isotype U).
Piper palustre C. DC. in A. DC., Prodr. 16(1): 293. 1869. Type locality: Brazil; type not designated).

Stout herb or slender shrub, ca. 1 m tall. Stem pubescent. Petiole 0-0.4 cm long, pubescent, vaginate at base; ligule present, 0.5-1 mm long, but often hidden by indument; blade scabrous, not conspicuously glandular-dotted, yellow-green, ovate, 5-12 x 2.5-6.5 cm, apex long-acuminate to obtuse, base equal or almost equally attached to petiole, unequally cordate, lepidote-scabrid above, lower surface and veins hirsute; pinnately veined, secondary veins 3-5 per side, originating from lower 1/3 of primary vein, impressed above, prominent below. Inflorescence erect; peduncle 0.2-0.6 cm long; spike 3.5-7 cm long, green, not apiculate; rachis glabrous or sparsely pubescent; floral bracts ciliate and hirtellous. Infructescence erect, 3.5-7 cm long; fruits globose to trigonous, 0.8-1 mm wide, glabrous, stigmas 3, sessile.

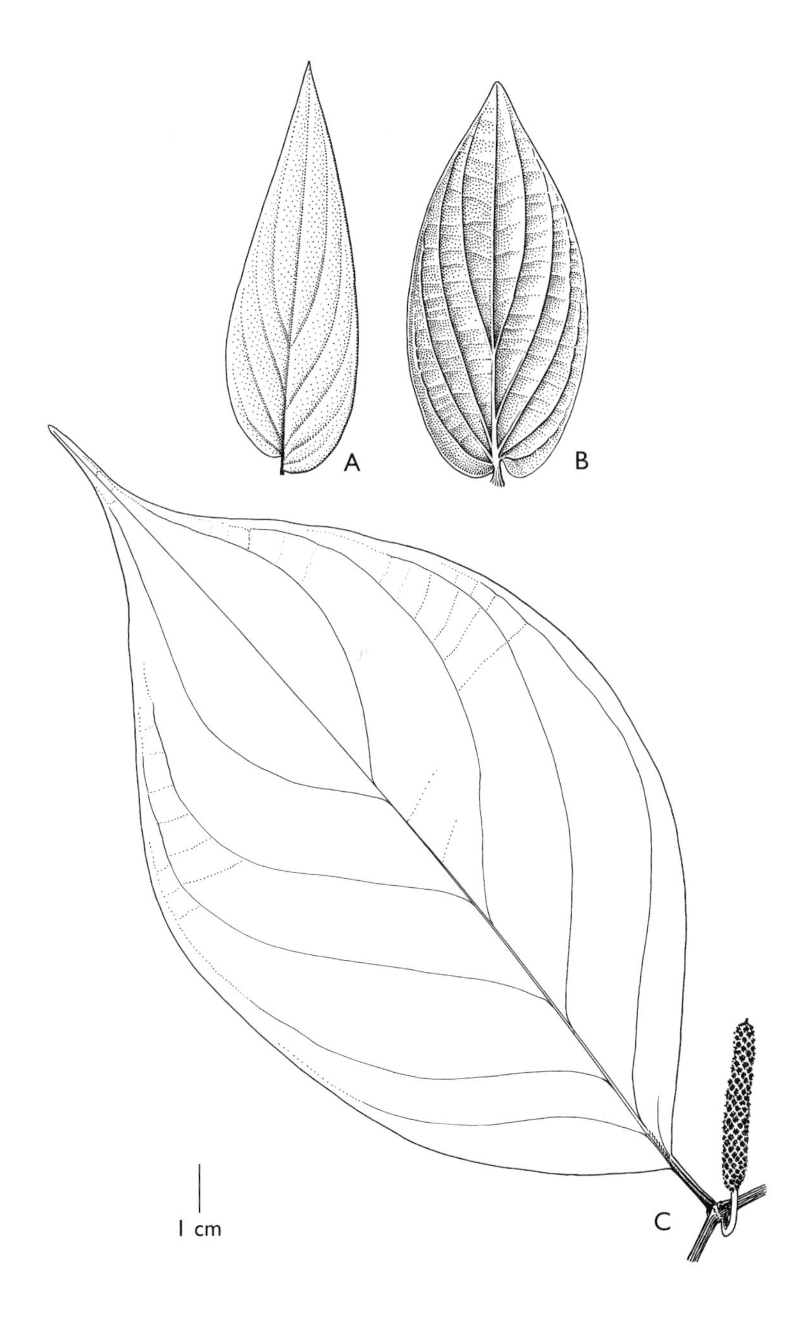

Fig. 39. *Piper fuligineum* (Kunth) Steud.: A-B, leaves. *Piper glabrescens* (Miq.) C. DC.: C, habit (A, Jansen-Jacobs *et al.* 2534; B, Oldenburger *et al.* 74; C, Hahn 4201).

Distribution: Guyana, Suriname, Brazil (Minas Gerais, Mato Grosso, Amazonas, Parana) and Paraguay; on edges of streams in savanna areas, in Brazil up to 1200 m elev.; 5 collections studied (GU: 3; SU: 2).

Specimens examined: Guyana: Rupununi Distr., Kuyuwini R., Kuyuwini landing, Jansen-Jacobs *et al.* 2534 (BRG, U); Takutu R., Gillespie 1915 (U); Makatui savanna, near Aishalton, Stoffers & Görts-van Rijn *et al.* 347 (NY, U). Suriname: Palaime savanna, Wessels Boer 843 (K, U); Sipaliwini savanna, Oldenburger e*t al.* 74 (U).

31. **Piper glabrescens** (Miq.) C. DC. in A. DC., Prodr. 16(1): 271. 1869. – *Artanthe glabrescens* Miq., London J. Bot. 4: 461. 1845. Type: Guyana, Parker s.n. (holotype K). – Fig. 39 C

Shrub to 1.5-2 m tall. Stem glabrescent. Petiole 1-3.2 cm long, vaginate to apex, glabrescent; blade not scabrous, densely glandular-dotted, elliptic to ovate, 14-28 x 7-14 cm, apex acuminate, base equal or unequally attached to petiole difference 0.2 cm, (acute to) rounded, glabrous above except minutely crisp-pubescent veins, glabrescent below; pinnately veined, secondary veins 5-7 per side, originating from lower 2/3 of primary vein, slightly impressed above, prominulous below, tertiary veins more or less parallel. Inflorescence erect; peduncle to 1.5 cm long, glabrous; spike to 6 cm long, apiculate; floral bracts glabrous to papillate. Infructescence to 0.5 cm thick; fruits depressed globose or subtetragonous, 1.2 mm in diam., glabrous, glandular, stigmas 3, sessile.

Distribution: Hispaniola, Puerto Rico, Trinidad, Venezuela, Guyana, French Guiana and Amazonian Brazil (Terr. Roraima); up to 900 m elev.; 35 collections studied (GU: 32; FG: 3).

Selected specimens: Guyana: Cuyuni-Mazaruni region, Paruima, McDowell 2638 (U); NW portion of Kanuku Mts., Iramaikpan, A.C. Smith 3663 (K, NY, U); Kanuku Mts., Jardin Falls, Jansen-Jacobs *et al.* 1228 (BBS, BRG, U). French Guiana: Mt. Galbao, de Granville *et al.* 8698 (CAY, U); trail to Boeuf Mort, Fournet 74 (CAY).

Notes: *Piper glabrescens* (Miq.) C. DC. var. *caparonum* (C. DC.) Yunck. from Trinidad and Guyana was described as having pilose floral bracts, but these were not found in the Guyanan collections. Therefore I do not accept this variety for the Guianas.
Piper pseudoglabrescens is differentiated from *P. glabrescens* by the measures of the leaves being more than 10 cm wide, instead of less than 10 cm wide in *P. pseudoglabrescens*. In the description, however, the

width is given as 7-14 cm. Another character used to differentiate the two taxa is the shape of the base of the leaf blade. But with the material at hand it is hard to follow the separation. One of the paratypes, A.C. Smith 3663, is studied, and it fits well in *P. glabrescens*, therefore *P. pseudoglabrescens* probably is a synonym of *P. glabrescens*.
For distinguishing characters with *Piper perstipulare* see note to the latter.

32. **Piper guianense** (Klotzsch) C. DC. in A. DC., Prodr. 16(1): 301. 1869. – *Artanthe guianensis* Klotzsch in Benth., J. Bot. (Hooker) 4: 322. 1841. Type: Brazil, Rio Branco, Ro. Schomburgk ser. I, 901 (holotype B not seen, isotypes BM, K, L, P). – Fig. 40 A

Artanthe schomburgkii Klotzsch in Benth., J. Bot. (Hooker) 4: 322. 1841. – *Artanthe peduncularis* Miq., Syst. Piperac. 531. 1844, nom. illeg. – *Piper pedunculare* C. DC. in A. DC., Prodr. 16(1): 291. 1869, nom. illeg. Type: Guyana, Curassawaka, Ro. Schomburgk ser. I, 696 (holotype G, not seen, isotypes, BM, K, P), syn. nov.
Artanthe oblongifolia Klotzsch in Benth., J. Bot. (Hooker) 4: 322. 1841. – *Piper oblongifolium* (Klotzsch) C. DC. in A. DC., Prodr. 16(1): 273. 1869. Type: Guyana, Parima Mts., Ro. Schomburgk s.n. Ao 1839 (holotype K).
Piper oblongifolium (Klotzsch) C. DC. var. *glabrum* C. DC. in A. DC., Prodr. 16(1): 273. 1869. Type: Guyana, Schomburgk s.n. (holotype G-DC, not seen), syn. nov.

Small shrub, often scrambling, rooting at lower nodes. Stem ribbed, glabrous, younger parts sparsely pubescent. Petiole 0.7-1.5 cm long, sparsely pubescent or glabrescent, vaginate to apex; blade not scabrous, somewhat glandular-dotted below, narrowly ovate to ovate, 5-8(-13) x 1.5-3.5(-6.3) cm, margin ciliate towards apex, apex acute to acuminate, base equal, obtuse, truncate, rounded or subacute, glabrous except sparsely minutely pubescent veins above, minutely crisp-pubescent on veins below; pinnately veined, secondary veins 4-5(-7) per side, originating from lower 1/3 or more of primary vein, 2 originating from base, tertiary venation reticulate obsolete above, slightly visible below, primary vein slightly impressed above, prominulous below. Inflorescence erect; peduncle filiform, to 3(-4) cm long, sparsely pubescent; spike 1-1.7 cm long, 1.5 mm wide, apiculate; floral bracts rounded to cucullate, sparsely fringed, with central dark gland; anthers dehiscing laterally; ovary glabrous. Fruits without style, stigmas 3?, sessile.

Distribution: Venezuela, Guyana, French Guiana, N Brazil and Peru; 6 collections studied (GU: 4; FG: 2).

Specimens examined: Guyana: Curassawaka, Ro. Schomburgk ser. I, 696 (BM. K, P).

Fig. 40. *Piper guianense* (Klotzsch) C. DC.: A, habit (A, Schomburgk 696).

Note: When Miquel in 1844 published *Artanthe peduncularis*, and as a synonym mentioned *A. schomburgkii* Klotzsch Mss., he did not realize that the latter had been published already in 1841, also described from Ro. Schomburgk ser. I, 696. I have studied the type collections of *A. schomburgkii* and compared them with the type collections and description of *P. guianense*. In my opinion the differences are too small to maintain *A. schomburgkii* (for which the combination in *Piper* never was published). Neither De Candolle nor Trelease and Yuncker give any additional collections, whereas *P. guianense* is reported from several parts of northern S America.

33. **Piper hirtilimbum** Trel. & Yunck., Piperac. N. South Amer. 291, f. 256. 1950. Type: Guyana, basin of Kuyuwini R., A.C. Smith 2632 (holotype US, isotype AA both not seen). – Fig. 41 A

Shrub to 0.5 m tall. Stem long-hirsute, hairs to 2.5 mm long. Petiole 1-3 cm long, long-hirsute, vaginate to apex (or to middle); blade not scabrous, glandular-dotted, obovate to elliptic or elliptic-oblong, 16-25 x 6-9.5 cm, margin long-ciliate, apex acuminate, base unequally attached to petiole difference 0-0.8 cm, rounded to subcordate, pilose above and below; pinnately veined, secondary veins 5-8 per side, originating from lower 2/3 of primary vein, flat above, prominent below, tertiary veins parallel and reticulate. Inflorescence pendent; peduncle 0.5-1.2 cm long, long-hirsute; spike 1-3 cm long, hardly apiculate; rachis presumably puberulent; floral bracts densely marginally fringed; ovaries (and fruits) pubescent, stigmas 3?, rounded, sessile.

Distribution: Guyana and French Guiana; in dense forest on brown sand, 240-700 m elev.; 12 collections studied (GU: 8; FG: 3).

Selected specimens: Guyana: Upper Takutu - Upper Essequibo region: Kuyuwini R. trail from river to Kassikaityu R., Clarke 4391 (U), Kamoa R., Clarke 3135 (U), Kamoa R., Toucan Mt., Jansen-Jacobs *et al.* 1749 (BRG, NY, U), Kuyuwini R., Kuyuwini Landing, Jansen-Jacobs *et al.* 3104 (BRG, U). French Guiana: Saül, Mt. Galbao, Boom 10783 (U); Grand Inini R., de Granville *et al.* C-73 (CAY, P, U).

Note: For differentiating characters with *Piper brownsbergense* see note to the latter.

34. **Piper hispidum** Sw., Prodr. 15. 1788. – *Piper hirsutum* Sw., Fl. Ind. Occid. 1: 60. 1797, nom. illeg. Type: Jamaica, Swartz s.n. (holotype S, not seen, isotype BM, not seen, B-W 666). – Fig. 41 B-D

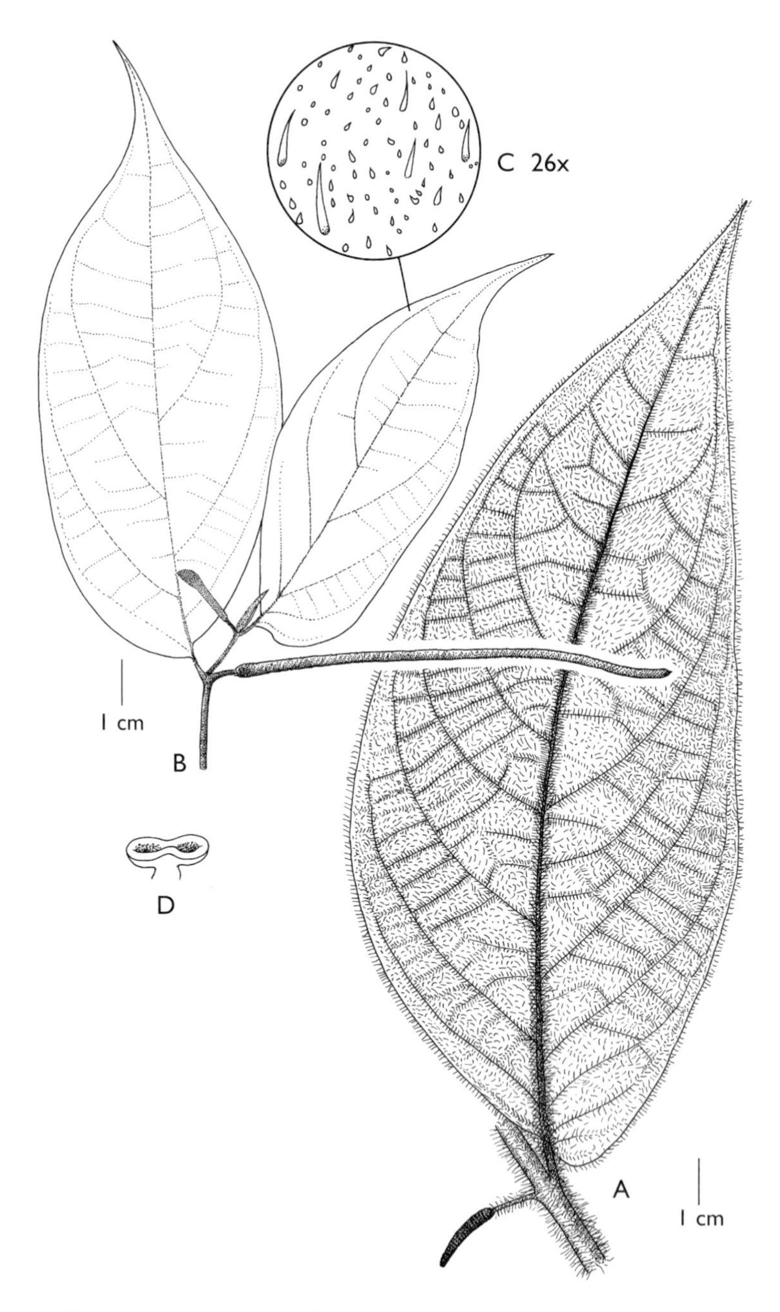

Fig. 41. *Piper hirtilimbum* Trel. & Yunck.: A, habit. *Piper hispidum* Sw.: B, habit; C, leaf indument; D, stamen (A, Clarke 4391; B-D, Hahn 3608).

Piper asperifolium Rich., Actes Soc. Hist. Nat. Paris 1: 105. 1792. Type: not designated.
Piper hispidum Sw. var. *magnifolium* C. DC. in A. DC., Prodr. 16(1): 276. 1869. Type: not designated.
Piper hispidum Sw. var. *obliquum* Trel. & Yunck., Piperac. N. South Amer. 273, f. 236. 1950. Type: Venezuela, lower Orinoco R., Rusby & Squires 58 (holotype US, not seen nor any of the many isotypes).
Piper trachydermum Trel., Contr. U.S. Natl. Herb. 26: 33. 1927. – *Piper hispidum* Sw. var. *trachydermum* (Trel.) Yunck., Ann. Missouri Bot. Gard. 37: 33. 1950. Type: Panama, S. Hayes 791 (holotype NY, not seen).

Shrub or subshrub, occasionally with scrambling branches, 2-3 m tall. Stem hirsute to glabrescent. Prophyll to 16 mm long, glabrous except pubescent on midvein. Petiole 0.5-1 cm long, hirsute; blade scabrous often on both surfaces, glandular-dotted, often greyish below, elliptic or elliptic-ovate, 11-23 x 4-11 cm, apex acuminate, base unequally attached to petiole difference 0.2-0.6 cm, obliquely rounded or cuneate with one side occasionally cordulate, lepidote-scabrid above, veins hirsute on both surfaces; pinnately veined, secondary veins 5-6 per side, originating from lower $^{1}/_{2}$ of primary vein, tertiary veins transversely reticulate. Inflorescence erect; peduncle to 1 cm long, pubescent;spike 8-14 cm long, white, cream or pale green, not apiculate; floral bracts densely marginally fringed, 0.4-0.5 mm in diam., filaments with broadened connective and anthers dehiscing transversely (horizontally) apically. Infructescence grey or grey green; fruits oblongoid to rounded-trigonous, hirsute, stigmas 3, sessile.

Distribution: West Indies, C and northern S America; in secondary vegetation, on creek banks in forest and in dense forest, from sea level to 700 m elev.; over 220 collections studied (GU: 80; SU: 52; FG: 91).

Selected specimens: Guyana: locality unknown, Harrison 1362 (BRG, K, NY); Rupununi Distr., Nappi head on Nappi R., Jansen-Jacobs *et al.* 615 (BBS, BRG, K, NY, P, U). Suriname: Upper Coppename R., Lindeman 6447 (BBS, NY, U); Wilhelmina Mts., Juliana peak, Irwin *et al.* 54654 (K, NY, U). French Guiana: Arataye, Parare Falls, Sastre 5656 (CAY, K, P, U); Nouragues, Cremers *et al.* 10887 (K. NY, P, U, US).

Vernacular names: Suriname: granman-ana-oedoe (Rombouts 629). French Guiana: granman-ana-oudou (BAFOG 7970); tubali (Arawak; Capus 151); pakilalu (Wayampi) = y m l u (Wayampi; Jacquemin 1516).

Notes: For distinguishing characters see note to *Piper dilatatum*. Yuncker (1957: 241-243), recognised four varieties; the differences are so small that I do not accept them here.

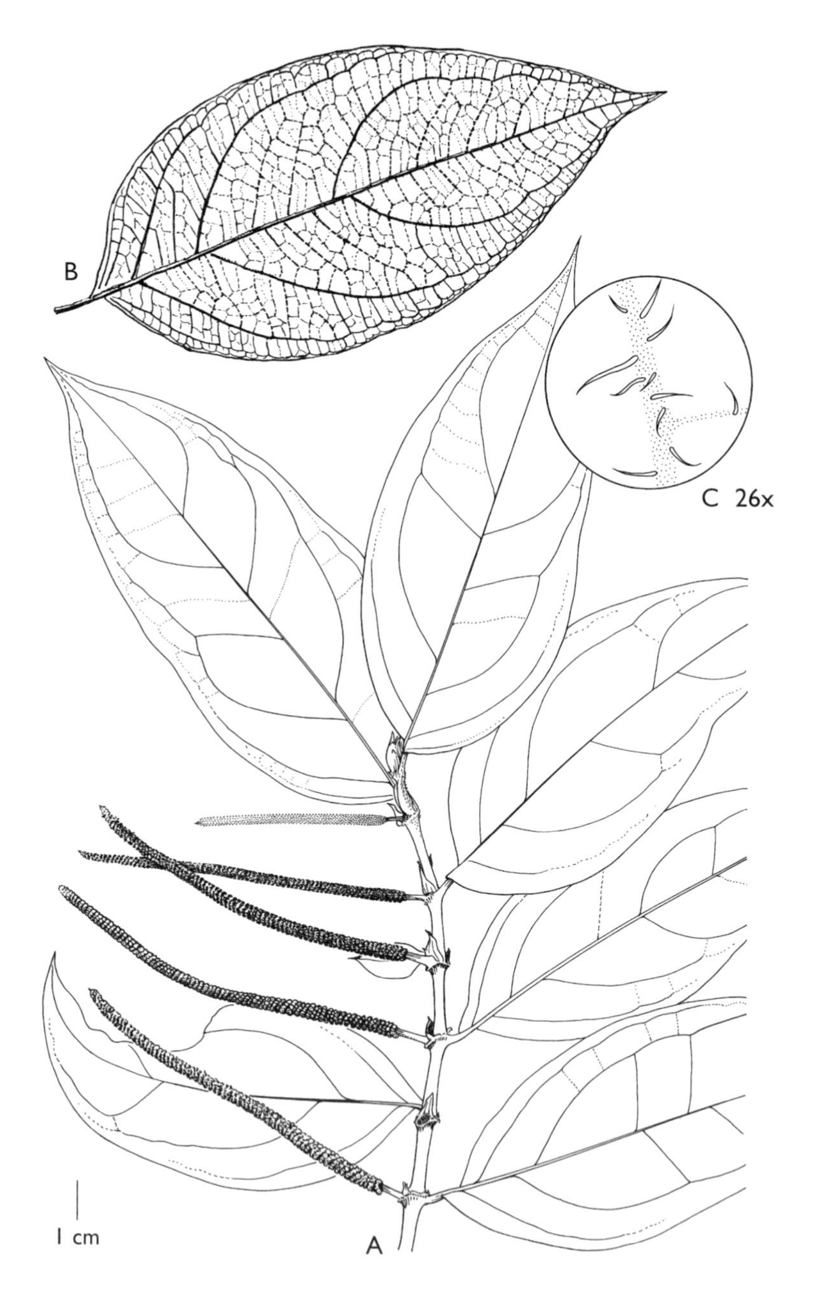

Fig. 42. *Piper hostmannianum* (Miq.) C. DC.: A, habit; B, leaf; C, indument of leaf lower surface (A, C, Mori *et al.* 23937; B, Hekking 1209). B, drawing by W.H.A. Hekking.

35. **Piper hostmannianum** (Miq.) C. DC. in A. DC., Prodr. 16(1): 287. 1869. – *Artanthe hostmanniana* Miq., London J. Bot. 4: 465. 1845. Type: Suriname, Hostmann 116 (holotype K). – Fig. 42 A-C

Piper hostmannianum (Miq.) C. DC. var. *ramiflorum* C. DC. in A. DC., Prodr. 16(1): 287.1869. Type: Suriname, Kappler 1397 (lectotype P, isolectotype U) (designated by Trel. & Yunck. 1950: 307).
Piper gleasonii Yunck. var. *wonotoboense* Yunck. in Trel. & Yunck., Piperac. N. South Amer. 364. 1950. Type: Suriname, Wonotobo, Stahel & Gonggrijp BW 2857 (holotype U) syn. nov.

Shrub, subshrub or treelet to 7 m tall. Stem crisp-pubescent. Prophyll glabrous, marginally with some hairs or densely pubescent. Stipules rather long persistent; petiole 0.5-1(-1.5) cm long, crisp-pubescent, vaginate to apex; blade not conspicuously glandular-dotted, shiny, broadly elliptic or broadly to narrowly ovate, 10-25 x 2.5-13 cm, (old leaves often broader than young ones), margin not ciliate, apex acuminate, base unequally attached to petiole difference 0-0.7 cm, obtuse, rounded to subcordate, sometimes acute, glabrous above, below densely appressed pubescent on veins; pinnately veined, secondary veins 3-6 per side, originating from lower 3/4 of primary vein, at an angle of 60°, anastomosing, soon abruptly ascending at an angle of 120°, intersecondaries conspicuous, tertiary veins widely reticulate. Inflorescence erect; peduncle 1 cm long to 1.5 cm in fruit, crisp-pubescent; spike 10-12 cm long, white or yellow; floral bracts densely crowded, spirally arranged, densely marginally fringed. Infructescence green, to 12 cm long, 4 mm wide; fruits oblongoid or trigonous, glabrous to puberulent at apex, stigmas 3, sessile.

Distribution: Northern S America; one of the most common species in the Guianas; over 250 collections studied (GU: 130; SU: 36; FG: 98).

Selected specimens: Guyana: Rupununi Distr., Kuyuwini Landing, Jansen-Jacobs *et al.* 2922 (BBS, BRG, K, P, U); Bartica-Potaro road, Sandwith 1153 (K, U). Suriname: Kabalebo dam area, Lindeman & Görts-van Rijn *et al.* 247 (BBS, K, NY, U); Man kaba, Sauvain 559 (BBS, P, U). French Guiana: Piste de Risque-tout, Skog 5619 (CAY, NY, P, U, US); Saül, Carbet Mais, Prance 28104 (CAY, K, NY).

Vernacular names: Suriname: wetu njaisa (Ndjuka; Sauvain 559); ampuku wi (Ndjuka; Sauvain 419). French Guiana: apoucou man limbe limbe (Serv. Forest. 4037); akamikini (Serv. Forest. 4041).

Notes: Seedlings often have a pale green band along the primary vein. After having studied many recent collections that show a great variation

in leaf shape, I cannot maintain a separate var. *ramiflorum*, based on leaf measures alone.
Piper hostmannianum can be distinguished from *P. trichoneuron* by indument characters. Hairs on veins below are appressed-pubescent in *P. hostmannianum* and crisp-pubescent in *P. trichoneuron.*

36. **Piper humistratum** Görts & K.U. Kramer, Proc. Kon. Ned. Akad. Wetensch. C 69: 273, t. 1. 1966. Type: Between Coppename R. and Emma Range, Wessels Boer 1469 (holotype U). – Fig. 43 A-C

Quebitea guianensis Aubl., Pl. Hist. Guiane 2: 838, Table Noms Latins 25; 4: t. 327, as 'guyannensis'. 1775 (non *Piper guianense* (Klotzsch) C. DC. 1869). Type: French Guiana, Aublet s.n. (holotype BM).

Creeping herb or subshrub to 0.35 m tall. Stem villous often with white hairs. Petiole 1-2(-5.5) cm long, villous; blade discolourous, not scabrous, not conspicuously glandular-dotted, elliptic to broadly elliptic, 9-15(-18) x 5-7(-8.5) cm, apex obtuse or acutish, base unequally attached to petiole difference 0.2-0.5 cm, obliquely rounded, glabrous above, veins brown-villous below; pinnately veined, secondary veins 3-5 per side, originating from lower 3/4 or more of primary vein, anastomosing, impressed above, prominent below like tertiary veins, those loosely reticulate. Inflorescence erect; peduncle 1.5-2.5 cm long, villous; spike 2-4 cm long, green or yellow or "rosish", not apiculate; floral bracts glabrous, slightly cucullate. Infructescence occasionally pendent; fruits oblongoid, sunken, glabrous, stigmas 4, sessile.

Distribution: The Guianas and Brazil (Terr. Roraima, Amapá); often on rocks and in humid places in primary forest, up to 800 m elev.; 51 collections studied (GU: 1; SU: 12; FG: 38).

Selected specimens: Guyana: NW Distr., Morucca R., Kabura Backdam, van Andel 2430 (U). Suriname: SE Suriname near Tapoc, Plotkin 610 (BBS); Tumac Humac Mts., Mt. Telouakem, Acevedo 5939 (U). French Guiana: Saül area, Freiberg 216 (B, CAY), Mori *et al.* 14909 (CAY, NY, P), de Granville *et al.* 11026 (CAY, NY, P, U, US).

Vernacular name: French Guiana: zaõ saapatu (Saramacca; Sauvain 39).

Note: I agree with Steyermark and Howard (in Howard, J. Arnold Arbor. 64: 282. 1983) that *Quebitea guianensis*, nicely depicted in Aublet, belongs to *Piper humistratum.*

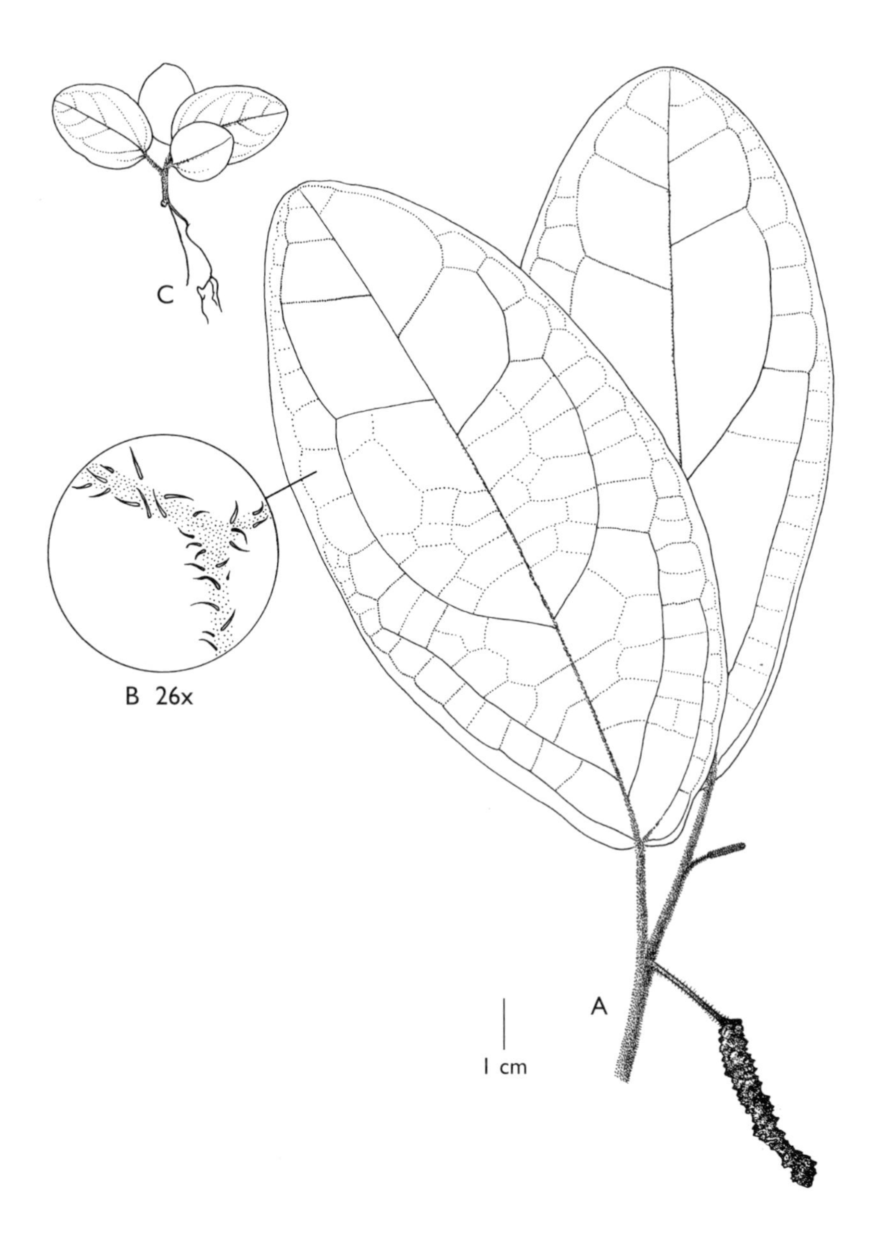

Fig. 43. *Piper humistratum* Görts & K.U. Kramer: A, habit; B, detail of vein indument; C, young plant (A-B, Acevedo 5939; C, Tjon Lim Sang 62).

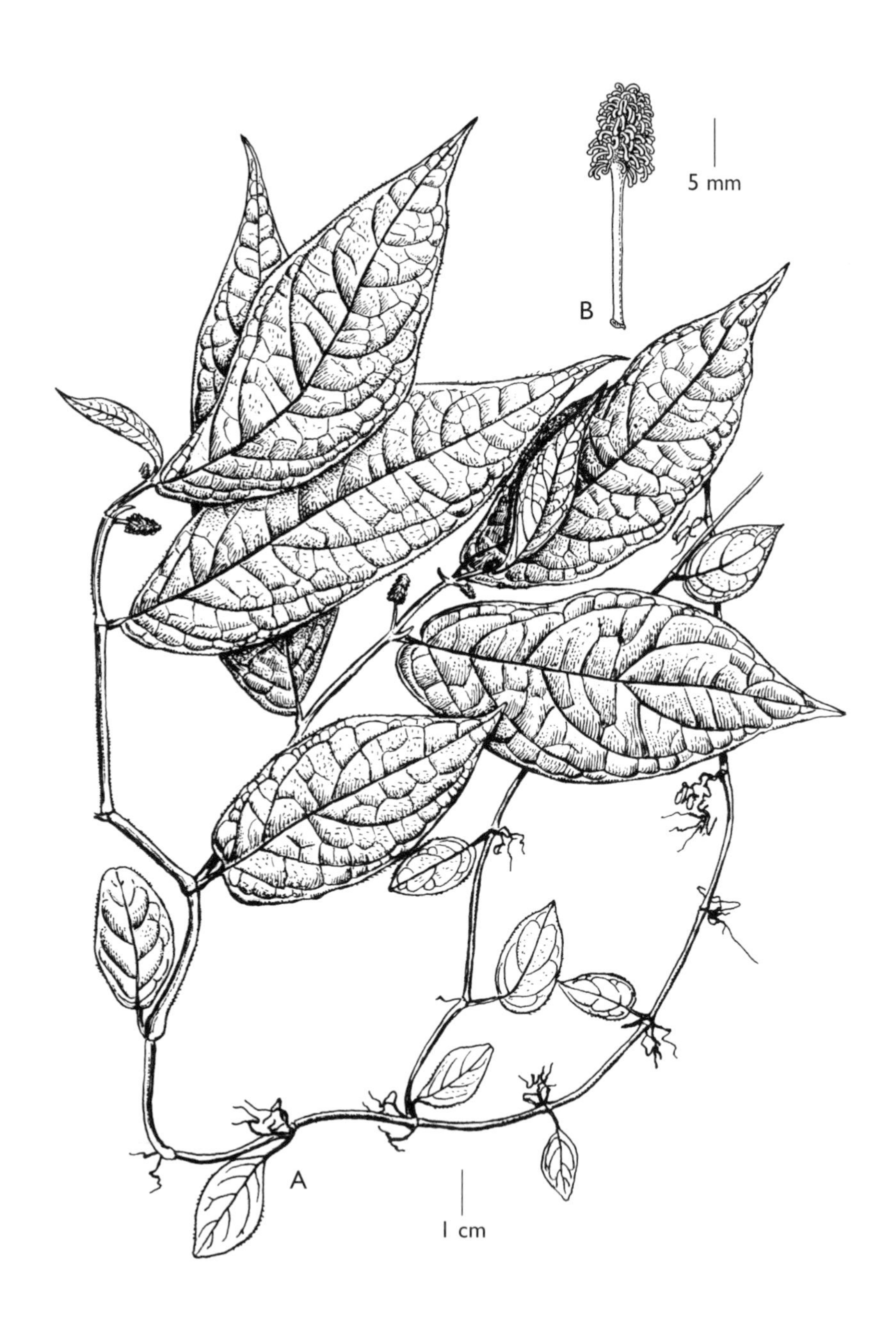

Fig. 44. *Piper hymenophyllum* (Miq.) Wight: A, habit; B, young flowering spike showing exserted recurved stamens (A, Kramer & Hekking 2666; B, Jenman 1851). A, drawing by W.H.A. Hekking.

37. **Piper hymenophyllum** (Miq.) Wight, Icon. Pl. Ind. Orient. 6: (4), t. 1942. 1853. – *Artanthe hymenophylla* Miq., Syst. Piperac. 532. 1844. – *Piper kegelii* C. DC. in A. DC., Prodr. 16(1): 323. 1869, nom. illeg. Type: Suriname, Hostmann 217 (holotype G-DEL, not seen, isotypes K, U). – Fig. 44 A-B

Nematanthera guianensis Miq., Linnaea 18: 607, f. 11. 1845 ('1844'). – *Piper nematanthera* C. DC. in A. DC., Prodr. 16(1): 367. 1869 (non *Piper guianense* (Klotzsch) C. DC. 1869). Type: Suriname, Para Distr., Kappler 1438 (holotype U, isotypes G-BOIS, P), syn. nov.
Piper sagotii C. DC., J. Bot. 4: 162. 1866, as 'sagoti'. – *Piper nematanthera* C. DC. var. *sagotii* (C. DC.) Trel. & Yunck., Piperac. N. South Amer. 378, f. 339. 1950. Type: French Guiana, Maroni, Sagot 1255 (holotype B, not seen, isotype K).

Scandent or creeping shrub, climbing branches rooting at nodes. Stem ridged, pubescent with recurved hairs. Petiole 0.5-1 cm long, pubescent, vaginate to apex; blade not conspicuously glandular-dotted, elliptic, subobovate or ovate, 4-17 x 3-6.5 cm, (on young creeping branches ca. 1.5-5 x 0.8-1.8 cm), margin occasionally ciliate, apex (long-)acuminate, base unequally attached to petiole difference 0.1-0.7 cm, unequally cordate or obtuse, glabrescent but veins may be recurved-pubescent; pinnately veined, secondary veins 6-8 per side, originating from throughout primary vein, impressed above, prominent below, tertiary veins inconspicuously reticulate. Inflorescence horizontally oriented? later recurved; peduncle 0.8-2 cm long, pubescent or glabrescent; spike protandrous (presenting staminate flowers first, pistillate flowers afterwards), 1-3.5 cm long, green, not apiculate; floral bracts pubescent at base, densely marginally fringed; stamen 1 per flower, far exserted, recurved, 2-3 mm long. Fruits obovoid to depressed globose, glabrous, stigmas 3 or 2, recurved, 0.5-1 mm long.

Distribution: The Guianas and Brazil; 34 collections studied (GU: 4; SU: 16; FG: 14).

Selected specimens: Guyana: Canje R., Jenman 1851 (K); Upper Takutu - Upper Essequibo region, Rewa R., Clarke 3951 (U). Suriname: Upper Nickerie R., camp Powiesi Boutoe, Jonker & Jonker-Verhoef 418 (U); Kamisa Falls, LBB 11057 (BBS, U); Upper Coppename R., Raleigh Falls, Voltzberg, Mennega 558 (U); Brownsberg, Görts-van Rijn *et al.* 505 (BBS, NY, U). French Guiana: Paul Isnard region, Feuillet 719 (CAY, U).

Notes: This creeping species is easily recognized by the difference in flowering pattern. At first sight the species seems to be dioecious, but actually it is protandrous. The flowers are unique in the Guianan species

in having only one stamen. The stamen has long filament that elongates before the ovary develops.
There is a great variety in leaf measurements, which may result in incorrect identification of immature collections.
When Wight used the name *Piper hymenophyllum*, illustrating his own specimen from India, he did not realize that this specimen did not belong to Miquel's species. Despite this, his new combination applies to the basionym, *Artanthe hymenophylla*.
De Candolle in transferring *Nematanthera guianensis* to *Piper nematanthera* cited two collections: Hostmann 19 (which is an error, because the number 10 is clearly written on the Kew voucher) and Kappler 1438. Miquel only mentioned the Kappler specimen which thus is the type of *Nematanthera guianensis*.
Trelease and Yuncker reduced *Piper sagotii* to variety *sagotii* under *P. nematanthera*. They indicated that additional material might show intermediate leaf shapes and measures. In Jonker & Jonker-Verhoef 418 there is a young, creeping branch bearing leaves of ca. 1.7 x 0.8 cm, the size of leaves in var. *sagotii*, whereas the leaves on the other branches are up to 17 x 6 cm, the size as described for the typical variety. To conclude, there is no reason to maintain var. *sagotii*.

38. **Piper inaequale** C. DC., J. Bot. 4: 217. 1866. Type: French Guiana, L.C. Richard s.n. (holotype P). – Fig. 45 A

Piper amapense Yunck., Brittonia 8: 234. 1957. Type: Brazil, Amapá, Oyapock Airfield, Cowan 38698 (holotype NY), syn. nov.

Shrub or treelet to 3(-5) m tall, almost glabrous. Petiole 2.5-4.5(-7.5) cm long, vaginate and narrowly winged to apex; blade sometimes discolorous and pale brown beneath, not scabrous, not conspicuously glandular-dotted, elliptic-oblong or somewhat ovate, 20-24(-36) x 7.5-9.5(-17) cm, apex acute to short-acuminate, lower half of blade asymmetrical, base unequally attached to petiole difference 0-0.5 cm: shorter side obtuse to abruptly acute, longer side abruptly rounded, obtuse or cordate, glabrous, except minutely pubescent veins below; pinnately veined, secondary veins 5-6(-8) per side, widely spaced, originating from lower 2/3 or more of primary vein, anastomosing well within margin, prominent below, tertiary veins very widely reticulate, minutely pubescent below. Inflorescence pendent; peduncle slender, 1.5-2.5 cm long, glabrous; spike 1.5-5 cm long, to 0.5 cm wide, not apiculate; rachis glabrous; floral bracts densely marginally fringed. Infructescence to 9 cm long; fruits obovoid or slightly flattened, ca. 2 mm wide, somewhat separate, glabrous or minutely papillate, stigmas 3-4, lanceolate, sessile.

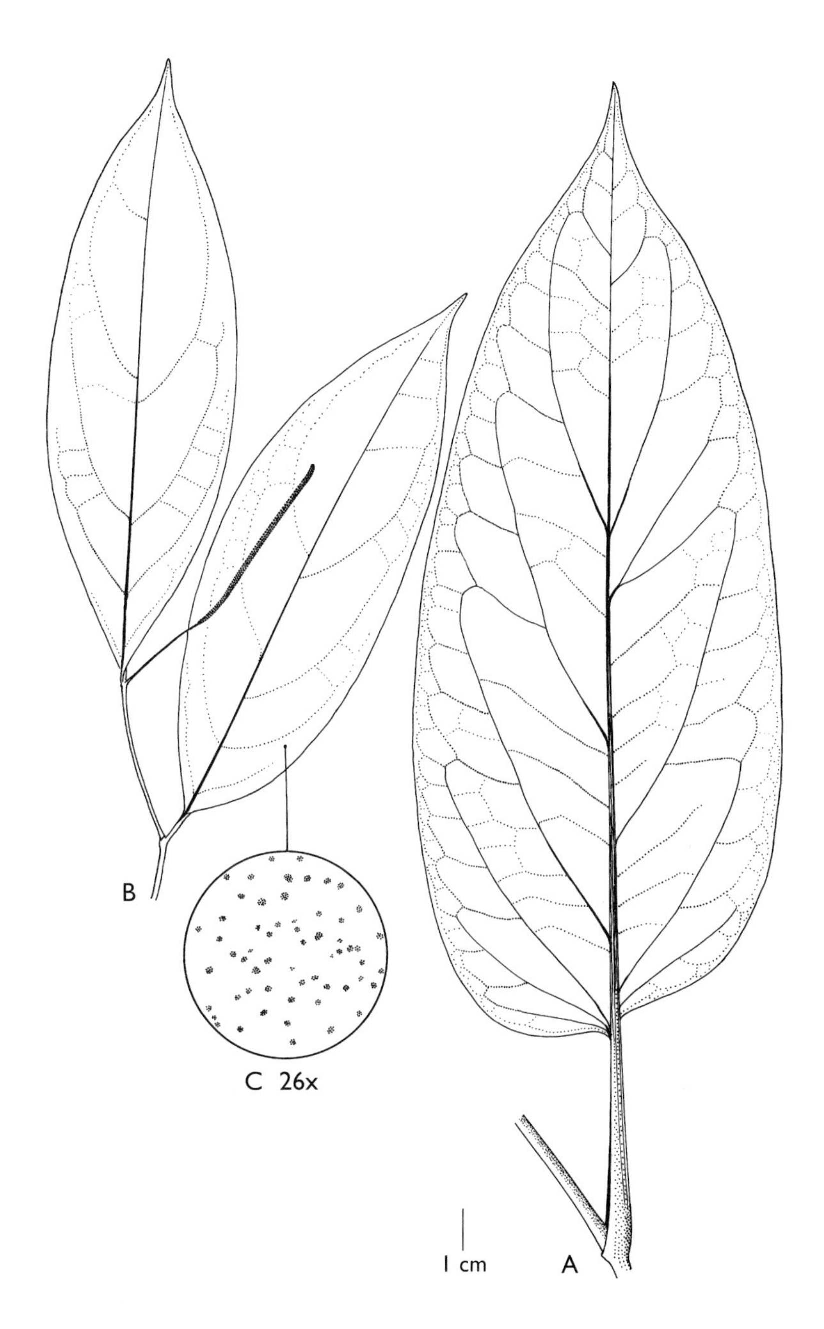

Fig. 45. *Piper inaequale* C. DC.: A, leaf. *Piper insipiens* Trel. & Yunck.: B, habit; C, detail of leaf lower surface with glandular dots (A, Irwin *et al.* 47720; B-C, de Granville *et al.* 11372).

Distribution: N Brazil (Amapá, Terr. Roraima) and French Guiana; in forest, to 150 m elev.; 35 collections studied (FG: 32).

Selected specimens: French Guiana: Upper Marouini R. basin, de Granville *et al.* 9953 (CAY, NY, P, U, US); Approuague R. basin, between Régina and Saint Georges, Cremers *et al.* 14245 (CAY, P, U); Oyapock R. area, Irwin *et al.* 47808 (NY, U), Oldeman 2727 (NY, P, U); Armontabo R., Prévost 1899 (CAY, NY, P, U); Zidock village, Prévost 1951 (CAY, NY, P, U).

Note: *Piper inaequale* up till now was an obscure taxon, known only from the type collection from French Guiana. It was not mentioned in the treatments neither of Steyermark nor of Burger. Trelease & Yuncker mentioned only the type collection, which is a specimen without fruits. After comparison of the type and further collections I conclude that *P. amapense* is conspecific with *P. inaequale*.

39. **Piper insipiens** Trel. & Yunck., Piperac. N. South Amer. 224. 1950. – *Artanthe glabella* Miq., Syst. Piperac. 518. 1844. – *Piper glabellum* (Miq.) C. DC. in A. DC., Prodr. 16(1): 312. 1869, non Swartz 1788. Type: French Guiana, Poiteau s.n. (holotype G-DEL, isotypes K, U). – Fig. 45 B-C

Shrub, subshrub or treelet, may be somewhat scandent, to 2 m tall. Stem, petiole and peduncle and even primary vein below maybe reddish (brown), or yellowish, glabrous. Stipules may be persistent, ca. 2 cm long, glabrous. Petiole 0.3-2.5 cm long, vaginate to apex; blade fleshy, drying membranous, discolorous often drying yellowish, conspicuously dark-glandular-dotted, elliptic, ovate, elliptic-oblong or rhombic, 13-23 x 4-8.5 cm, apex (long-)acuminate, base equal or unequally attached to petiole difference up to 0.5 cm, acute, cuneate or obtusish, glabrous above, glabrous often minutely crisp-pubescent below especially on veins; pinnately veined, secondary veins 4-6 per side, originating from lower $^1/_2$ or $^2/_3$ of primary vein, plane above, prominulous to prominent below, tertiary veins widely reticulate, areoles almost rectangular. Inflorescence erect; peduncle slender, 1-3.5 cm long; spike 7-13 cm long, white, yellowish or greyish-green, apiculate; rachis papillose; floral bracts triangular, densely marginally fringed. Infructescence green, erect or pendent; fruits trigonous, glabrous, yellowish to green, stigmas 3, short, sessile.

Distribution: The Guianas; dark, humid, primary forest or swamp forest, from sea level to 700 m elev.; ca. 85 collections studied (GU: 35; SU: 1; FG: 48).

Selected specimens: Guyana: Cuyuni-Mazaruni region, Hahn 5334 (U. US); Barima-Waini region, Pipoly 8208 (U, US). Suriname: Emma range, Daniëls & Jonker 1056 (U). French Guiana: Kaw Mts., Cremers *et al.* 9328 (CAY, NY, P, U); Trinité Mts., de Granville *et al.* 6561 (BR, CAY, G, MG, P, U); Sinnamary R. basin, saut Dalles, Hoff 6707 (CAY, NY, P, U).

40. **Piper marginatum** Jacq., Collectanea 4: 128. 1791. Type locality: Venezuela, Caracas; type not designated. – Fig. 46 A-B

Piper catalpifolium Kunth in Humb., Bonpl. & Kunth, Nov. Gen. Sp. ed. qu. 1: 58. 1816, as 'catalpaefolium'; – *Piper marginatum* Jacq. var. *catalpifolium* (Kunth) C. DC. in A. DC., Prodr. 16(1): 246. 1869, as 'catalpaefolium'. – *Piper marginatum* Jacq. var. *marginatum* f. *catalpifolium* (Kunth) Steyerm., Fl. Venez. 2(2): 480, f. 69. 1984, as 'catalpaefolium'. Type: Venezuela, Prov. Sucre, near Cumanacoa, Humboldt & Bonpland s.n. (isotype BW 699).
Piper anisatum Kunth in Humb., Bonpl. & Kunth, Nov. Gen. Sp. ed. qu. 1: 58. 1816. – *Piper marginatum* Jacq. var. *anisatum* (Kunth) C. DC. in Urb., Symb. Antill. 3: 172. 1902. Type: Venezuela, Orinoco R., near Hato del Capuchino, Humboldt & Bonpland 1056 (holotype P, isotype B-W 698, none seen).
Piper marginatum Jacq. var. *clausum* Yunck., Ann. Missouri Bot. Gard. 37: 17. 1950. Type: Panama, Canal Zone, Miller 1861 (holotype US, not seen).

Subshrub or treelet, to 3 m tall; fragrant like anise. Petiole vaginate to apex, 4 cm long, glabrous; blade not dark glandular-dotted, rounded to ovate, 10-12 x 8-15 cm, margin ciliate, apex (long-)acuminate, base equal or almost equally attached to petiole, deeply cordate, glabrescent or veins puberulent; palmately 9-11-veined. Inflorescence erect and recurved; peduncle 1 cm long, glabrous; spike 10-13(-20) cm long; floral bracts densely marginally fringed. Fruits oblongoid, glabrous, stigmas 3, sessile.

Distribution: From Mexico south to Brazil and Ecuador, the West Indies; over 100 collections studied (GU: 16; SU: 48; FG: 37).

Selected specimens: Guyana: Upper Takutu - Upper Essequibo region, Surana village, McDowell 1903 (U); Kanuku Mts., Nappi-head on Nappi Cr., Camp 1, Jansen-Jacobs *et al.* 617 (BBS, K, U). Suriname: Tapanahoni R., Tabiki, Sauvain 387 (BBS, P, U); Maroni R., Sastre 8187 (CAY, P, U). French Guiana: Sinnamary, Lescure 227 (CAY, P, U); Cayenne, Prévost 3633 (CAY, NY, P, U).

Note: When Kunth described *Piper catalpifolium*, he already observed that it is related to *P. marginatum*. Steyermark accepted this taxon as *P. marginatum* var. *marginatum* f. *catalpifolium*. The differences are found

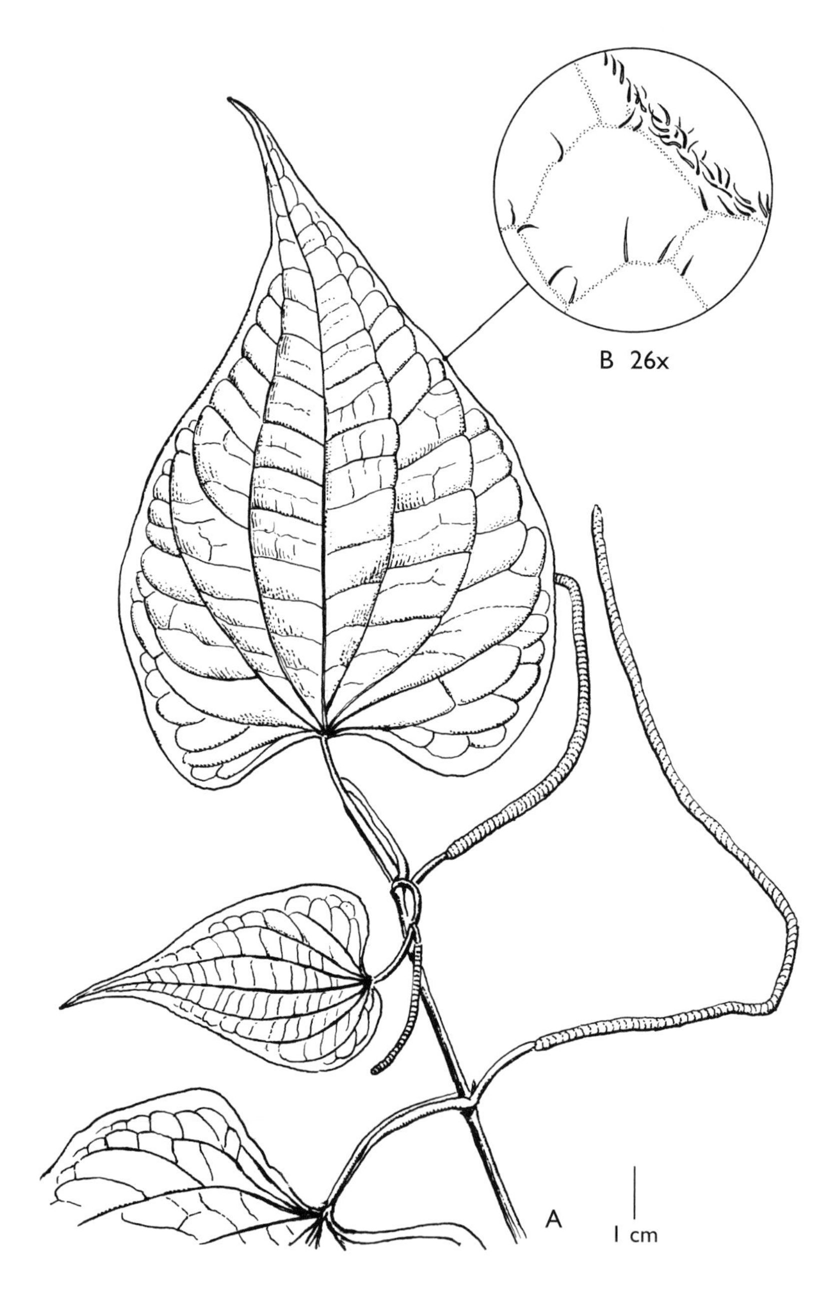

Fig. 46. *Piper marginatum* Jacq.: A, habit; B, detail of leaf margin (A-B, Henkel 3910).

in the leaf indument consisting of more or fewer hairs on the leaf upper surface. To have to look for some hairs or a few more hairs on the leaves is no argument to create infraspecific taxa. Therefore I do not accept this taxon for the Guianas.
Yuncker (1957: 230-231), recognised two more varieties: var. *anisatum* and var. *clausum*; the differences are so small that I do not accept them here.
Piper marginatum is easily recognizable by the deeply cordate, palmately veined leaves.

41. **Piper nigrispicum** C. DC. in J. Huber, Bol. Mus. Goeldi Hist. Nat. Ethnogr. 5: 330. 1909. Type: Brazil, Pará, Alto Ariramba region, porto do Jaramacarú, Ducke 8059 (holotype MG, isotype G, both not seen). – Fig. 47 A-B

Shrub to 2.5 m tall. Stem glabrous, young ones often reddish. Petiole 0.4-0.8 cm long, glabrous, pubescent when young, vaginate at base; blade shiny dark green above, pale below, densely black glandular-dotted, translucent dotted *in vivo,* (narrowly) ovate to elliptic-oblong, 8-15 x 3-6 cm, apex acuminate, base equal or almost equally attached to petiole, acute or cuneate or obtusish, occasionally rough above (from scale-like emergent structures), glabrous below; pinnately veined, secondary veins 3-5 per side, anastomosing near margin, originating from lower 3/4 of primary vein, tertiary veins widely reticulate, not conspicuous. Inflorescence pendent; peduncle slender, 0.5-1.8 cm long, glabrous, often reddish or violaceous; spike 1-5 cm long, 10 mm in diam. *in vivo*, green, yellow to reddish, apiculate; floral bracts densely marginally fringed. Infructescence pendent, green; fruits trigonous or subtetragonous with narrow "collar", glabrous, green, stigmas 3, sessile.

Distribution: NE Brazil and the Guianas (mainly French Guiana) and Amazonian Bolivia; in forest from sea level to 700 m elev.; over 120 collections studied (GU: 6; SU: 1; FG: 114).

Selected specimens: Guyana: Mabura, Ek 551 (U); NW Distr., Barima R., Kariako, van Andel 1469 (U). Suriname: Tumac Humac Range, Telouakem, Acevedo 6019 (U, US). French Guiana: Saül, Görts-van Rijn *et al.* 53; Mt. Galbao, de Granville *et al.* 9006 (B, CAY, MG, MO, NY, P, U, US); Mt. Bellevue de l'Inini, de Granville *et al.* 7579 (B, BR, CAY, MG, MO, P, U).

Note: Many collections were previously assigned to *Piper rudgeanum.* After having compared those collections with the type of *P. rudgeanum* (Suriname, Kegel s.n.), it is certain that they belong to *P. nigrispicum. Piper rudgeanum* differs in having 1 cm long spikes on stout peduncles,

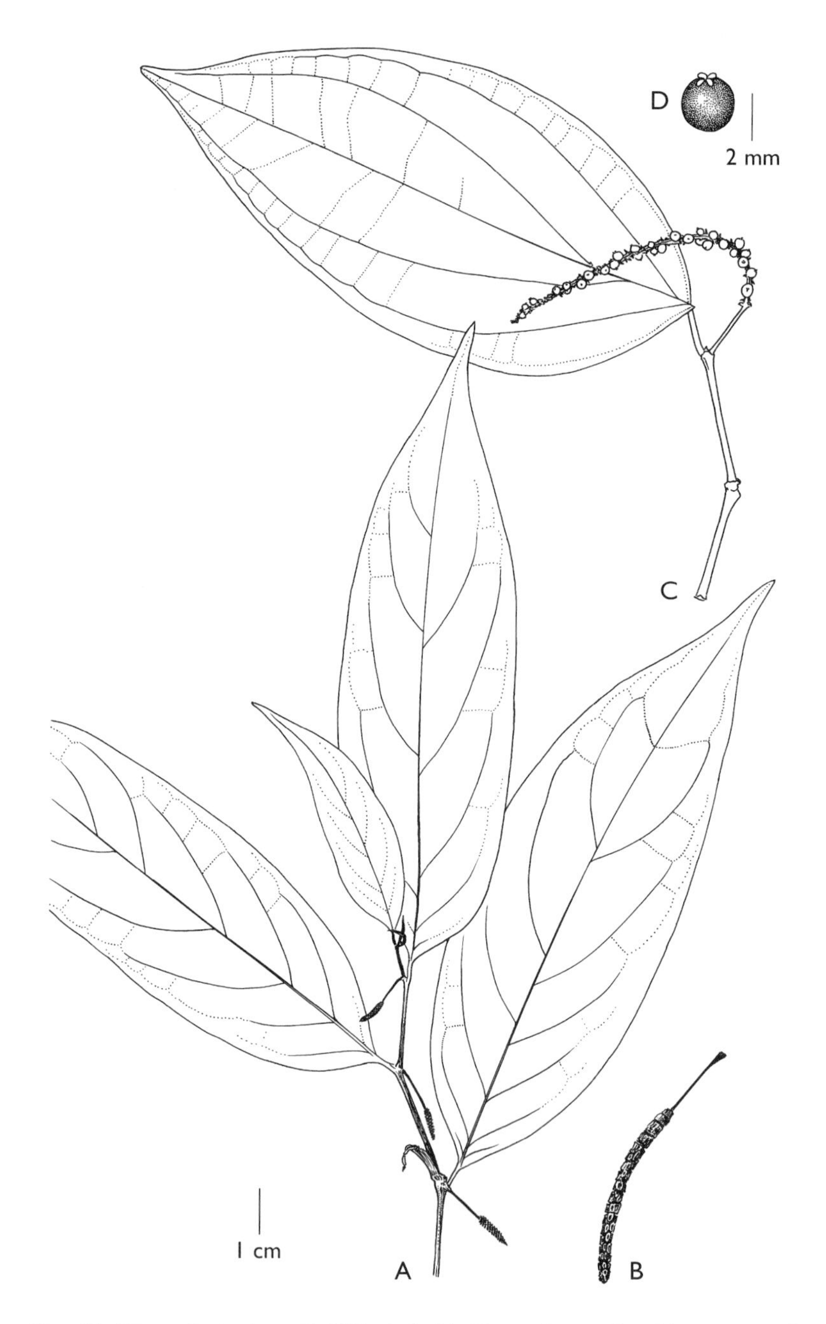

Fig. 47. *Piper nigrispicum* C. DC.: A, habit; B, pendent spike. *Piper nigrum* L.: C, habit; D, fruit (A-B, de Granville *et al.* 7579; C-D, Herbarium Lemée s.n.).

stylose ovaries and fruits subtended by cucullate floral bracts with only the ventral margin fringed, whereas in *P. nigrispicum* slender peduncles bear 1-5 cm long spikes and the ovaries and fruits have sessile stigmas and are subtended by densely marginally fringed floral bracts.

42. **Piper nigrum** L., Sp. Pl. 28. 1753. Type locality: India; type not designated. – Fig. 47 C-D

Shrub sometimes with scrambling branches or liana. Stem glabrous. Petiole 1-2 cm long, glabrous, vaginate to apex; blade not scabrous, not glandular-dotted, elliptic, ovate to broadly ovate, 8-19 x 5-10 cm, apex acute to short-acuminate, base equal or almost equally attached to petiole, subacute to obtusish, glabrous; palmately-pinnately veined, secondary veins 2-3 per side, originating from lower 1/4 of primary vein, flat to prominulous above, prominulous to prominent below, tertiary veins reticulate. Inflorescence pendent; peduncle 1-2.5 cm long; spike 5-10 cm long, apiculate or not; floral bracts glabrous. Infructescence pendent, 12 cm long, more than 0.5 cm thick. Fruits globose, 5-6 mm long, glabrous, red or black, stigmas 2-3, sessile.

Distribution: Origin W India; cultivated in Brazil (introduced in 1930) and the West Indies; 10 collections studied (SU: 1; FG: 9).

Selected specimens: Suriname: without locality, UVS (Kaipoe) s.n. (BBS). French Guiana: Station de recherche de Paracou, Bordenave 63 (U); Abandoned dwelling, Lemée s.n. (P); between Kourou and Pariacabo, Sagot s.n. (P).

Vernacular name: pepper.

Use: Pepper is the most used spice in the world, although it contains several contents causing hypoglycaemic effects in humans. To obtain "black pepper" the unripe fruits are collected. During the drying proces its colour changes from green to black. To produce "white pepper" fruits are allowed to mature further, then soaked in (streaming) water. The mesocarp is then removed. In Suriname powder of pepper plus salt and lemon juice is said to alleviate coughs by East Indian people (Raghoenandan, Internal report BBS).

Notes: Hardly any recent collections of pepper have been reported.

Two other spicy pipers are mentioned below. No collections have been reported. They are certainly known and used in the Guianas:

Piper betle L.

Uses: Cultivated in the Indomalesian region (widely cultivated already since at least Hobhinian culture 8000-3000 years ago). In Suriname crushed leaves are used to stop nose bleeding. Other medicinal uses as known from India have not been reported (Raghoenandan, Internal report BBS, on medicinal uses of plants by East Indian people in Suriname). Fresh leaves made up into a packet or plug containing betel nuts (*Areca catechu)* and slaked lime provide a masticatory.

Piper cubeba L.f.

Uses: Cultivated in E and W Indies for the dried unripe fruits used medicinally and to flavour cigarettes.

43. **Piper obliquum** Ruiz & Pav., Fl. Peruv. Chil. 1: 37, t. 63, f. a. 1798. Type: Peru, Huanuco, Cuchero near Cayumba, Ruiz & Pavón s.n. (holotype P; isotype BM?, photo F, none seen). – Fig. 48 A-B

Piper caracasanum Bredem. in Link, Jahrb. Gewächsk. 1(3): 61. 1820. Type: Venezuela, Caracas, Bredemeyer s.n. (holotype B-W 679, not seen).
Steffensia insignis Kunth, Linnaea 13. 667. 1840 ('1839'). – *Piper insigne* (Kunth) Steud., Nomencl. Bot. ed. 2. 2: 341. 1841. – *Artanthe insignis* (Kunth) Miq., Syst. Piperac. 394. 1844. Type: French Guiana, near Cayenne, J. Martin s.n. (isotypes P, U), syn. nov.
Piper submelanostictum C. DC., Notizbl. Bot. Gart. Berlin-Dahlem 7(62): 443. 1917. Type: French Guiana, Leprieur s.n. (holotype B, not seen).
Piper submelanostictum C. DC. var. *amelanostictum* Yunck. in Maguire *et al.*, Bull. Torrey Bot. Club 75: 289. 1948. Type: Guyana, Rockstone, Gleason 566 (holotype NY, not seen).
Piper saramaccanum Yunck. in Maguire *et al.*, Bull. Torrey Bot. Club 75: 288. 1948. Type: Suriname, near Krappa camp, Saramacca R. Headwaters, Maguire 24887 (holotype NY, isotype U).
Piper divulgatum Trel. & Yunck., Piperac. N. South Amer. 153, f. 130. 1950. Type: Colombia, Norte de Santander, Cuatrecasas 12964 (holotype US).

Shrub or treelet, 1.5-8 m tall. Stem densely pubescent to glabrescent. Petiole 3-7(-10) cm long, glabrous or densely brown-pubescent, green to brown, vaginate or winged to apex; blade not scabrous, not glandular-dotted, elliptic-ovate to oblong, 20-60(-70) x (12-)20-35 cm, apex acute to acuminate, base unequally attached to petiole difference up to 1 cm, unequally lobed, basal lobes shorter or longer than petiole, occasionally overlapping the latter, glabrous above, more or less brown-pubescent below, especially on veins; pinnately veined, secondary veins 5-9, per side originating from lower 3/4 of primary vein. Inflorescence pendent; peduncle 2-5 cm long, pubescent to glabrescent; spike (8-)20-60(-70) cm

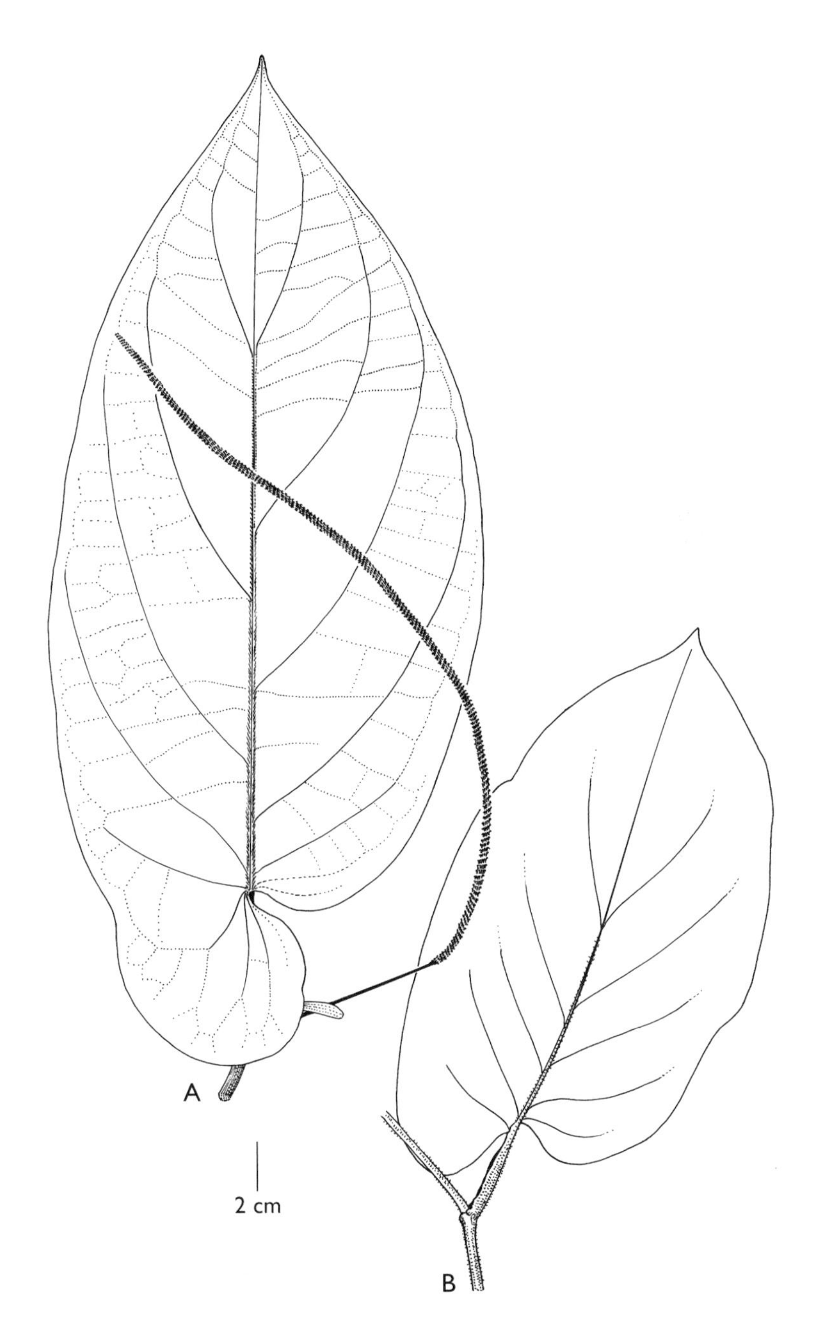

Fig. 48. *Piper obliquum* Ruiz & Pav.: A, habit; B, leaf (A, Sastre 8066; B, Hahn 5701).

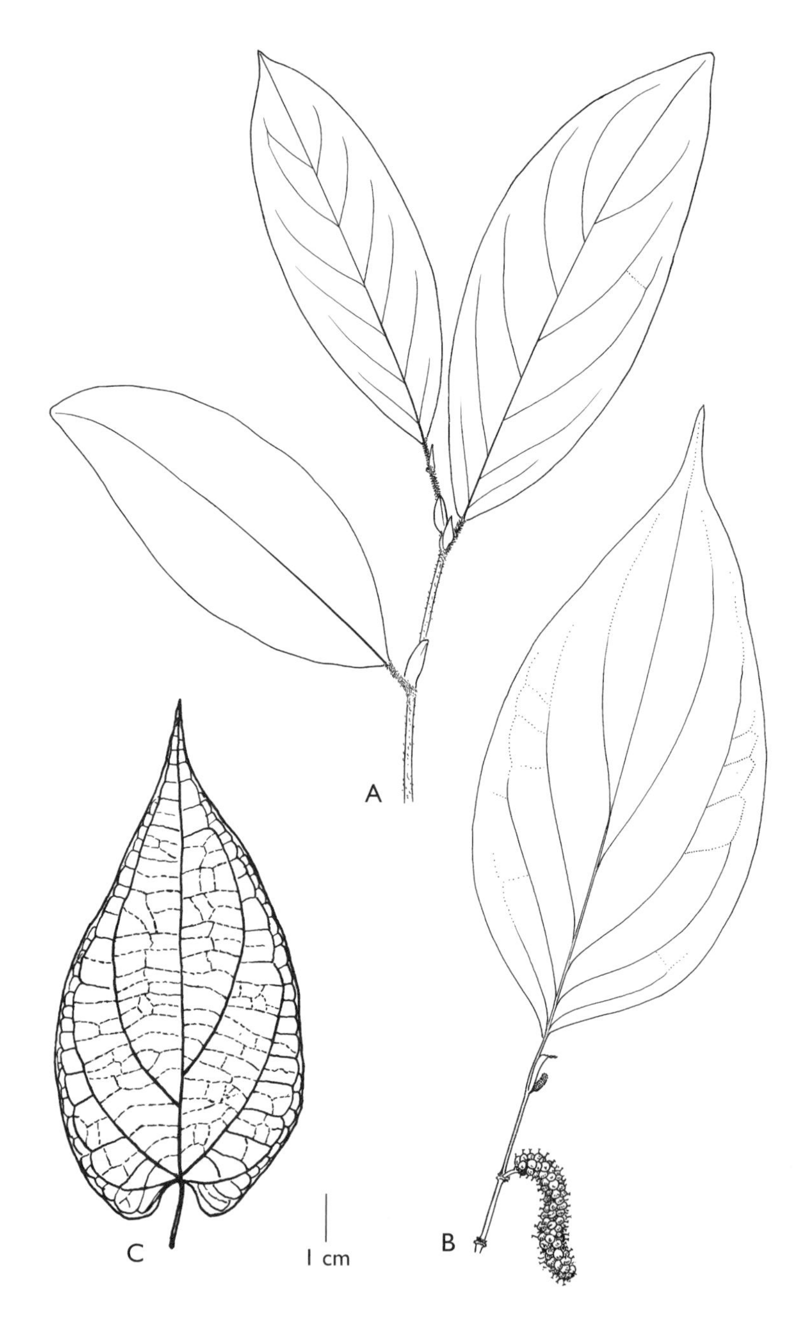

Fig. 49. *Piper paramaribense* C. DC.: A, habit. *Piper phytolaccifolium* Opiz: B, habit. *Piper poiteanum* Steud.: C, leaf (A, de Granville *et al.* 7864; B, de Granville *et al.* 8634; C, Happle s.n.). C, drawing by W.H.A. Hekking.

long, reddish when young, apiculate; rachis pubescent; floral bracts triangular-rounded to cucullate, glabrous. Fruits obovoid or oblongoid, sometimes flattened at apex, 1-2 mm in diam., minutely pubescent or glabrous or with a few basal hairs, stigmas 3, often quite long, sessile, sometimes on a short style.

Distribution: Amazonian Brazil, Peru and the Guianas; in moist forest from sea-level to 2000 m elev.; ca. 90 collections studied (GU: 23; SU: 10; FG: 57).

Selected specimens: Guyana: Pakaraima Mts., N of Paruima, Maas *et al.* 5856 (BBS, K, P, U); Cuyuni-Mazaruni region, Pipoly 11015 (BRG, NY, U, US). Suriname: Wilhelmina Mts., Irwin *et al.* 54759 (NY, U); Emma Range, Daniëls & Jonker 1254 (NY, U). French Guiana: Trinité Mts., de Granville *et al.* 6410 (CAY, K, NY, P, U, US); Upper Marouini basin, de Granville *et al.* 9223 (CAY, NY, P, U, US).

Vernacular names: French Guiana/Brazilian border: Galet fin de Rio Jari (Lescure 314); yalataku asili (Wayampi).

Note: There is a great variation in leaf shape and size and in the presence and density of the indument of leaf and stem. This has resulted in the creation of many species. Tebbs (1989: 129, f. 15, 17), in studying the variation of leaves in the taxon, concluded that many names can be reduced to synonymy. In her key, Tebbs separated *Piper obliquum* and the other large-leaved, tree-like *P. cernuum* by the indument of the fruits.

44. **Piper paramaribense** C. DC. in A. DC., Prodr. 16(1): 297. 1869. Type: Suriname, Paramaribo, Wullschlaegel 1563 (holotype BR, isotype BR, both not seen). – Fig. 49 A

Piper maguirei Yunck. in Maguire *et al.*, Bull. Torrey Bot. Club 75: 289. 1948. Type: Suriname, Tafelberg, lower North Ridge Cr., Maguire 24811 (holotype NY, isotypes K, U), syn. nov.
Piper fockei Trel & Yunck., Piperac. N. South Amer. 1: 355, f. 322. 1950. Type: Suriname, Focke 1234 (holotype K, isotype U), syn. nov.

Small shrub, sometimes trailing, to 1 m tall. Stem hirsute. Stipules long persistent. Petiole 0.5 cm long, hirsute, vaginate to middle; blade slightly bullate to smooth, coriaceous and silvery shiny or membranous, glandular-dotted, lance-elliptic, elliptic or elliptic-oblong, 7-17 x 2.5-6 cm, apex short-acuminate, maybe somewhat obtuse, base equal or unequally attached to petiole difference at most 0.1 cm, cuneate to obtusish, rounded or subcordate, glabrous but veins appressed-pubescent below; pinnately veined, secondary veins (7-)8-12 per side, originating from throughout primary vein, impressed above, prominulous to strongly prominent below, tertiary veins transverse. Inflorescence erect; peduncle 0.5-1 cm long, glabrous or hirsute; spike 2-4 cm long, yellow or green, apiculate; floral bracts cucullate and pilose on inner side and at base, rachis pubescent. Infructescence erect, to 7 cm long, to 1 cm thick, green; fruits globose or obovoid, 3 mm in diam., glabrous, green, stigmas 3, linear, erect, may give impression of a style, obsolete on fruit.

Distribution: The Guianas; up to 1600 m elev.; ca. 55 collections studied (GU: 2; SU: 12; FG: 43).

Selected specimens: Guyana: Cuyuni-Mazaruni region, Hahn 5318 (U); Upper Takutu-Upper Essequibo region, Acarai Mts., Clarke 7546 (U). Suriname: Kabalebo Dam area, Lindeman & Görts-van Rijn *et al.* 307 (BBS, U); Wilhelmina Mts. near Kayser airstrip, Irwin *et al.* 57710 (NY, U). French Guiana: Piste de St. Elie, Prévost 1168 bis (CAY, P, U); Saül, Circuit la Fumée, Prévost 1799 (CAY, U); Upper Marouini basin, de Granville *et al.* 9684 (CAY, P, U, US).

Notes: *Piper maguirei* has been described from a young specimen having membranous leaves, whereas *P. paramaribense* has coriaceous leaves. Comparing the type collection and the descriptions of *P. maguirei* and of *P. paramaribense* there seems to be no reason to keep the two taxa separate. The texture of the leaves, however, can not be maintained as a differentiating character when we find in Irwin *et al.* 55853 similar thin leaves together with characteristic coriaceous ones.
Piper fockei was described from a specimen that C. DC. had included in *P. hostmannianum* var. *berbicense*; C. DC. had misread Focke's name as Forbes. Comparing the type and other *P. fockei* collections with those of *P. paramaribense* and checking the descriptions I could see no distinguishing characters and decided to synonimize *P. fockei* under *P. paramaribense.* The protologue of the latter was somewhat incomplete and has been completed with the material now available. The plants are easily recognisable by the long-persistent stipules, silvery shiny leaves (in dried state) with strongly prominent veins below, and slightly stylose ovaries.
Piper fulgidum Yunck. from Amapá, Brazil is similar to *P. paramaribense.* I have not yet studied the type collection of the former; I suppose, however, that they are conspecific.
Yuncker (1957: 259) considered *P. pertinax* Trel. & Yunck. as a synonym of *P. paramaribense*. Because *P. pertinax* is a nom. nov. for *P. affine* (Miq.) C. DC. 1869, non *P. affine* M. Martens & Galeotti 1843, its type is the type of the basionym, *Artanthe affinis* Miq. The type of the latter, however, is not known. In Miquel's series of publications on Focke specimens from Suriname (Linnaea 18: 49-95, 225-240. 1845, etc.) he does not mention any Focke number. Yuncker (l.c. 260) lists under *P. paramaribense* a Focke

710 specimen from Vredenburger Zandrits, the locality that Miquel mentioned. Focke 710 in U, however, belongs to *P. avellanum*.

45. **Piper peltatum** L., Sp. Pl. 30, [1231]. 1753. – *Lepianthes peltata* (L.) R.A. Howard, J. Arnold Arbor. 54: 381. 1973. – *Pothomorphe peltata* (L.) Miq., Comm. Phytogr. 37. 1840. Type: [icon] Plumier, Descr. Pl. Amér. t. 74. 1693 (according to Howard 1973: 381).
– Fig. 50 A

Large herb to 3 m tall. Internodes black glandular-dotted, more densely so on nodes. Leaves alternate, peltate; petiole attached at up to 1/3 of blade, rarely near base, 8-20 cm long, vaginate or slightly winged in lower part; blade glandular-dotted, round-ovate, to 16-30 x 18-40 cm, apex acute, base nearly rounded to usually deeply cordate, glabrous except for short hairs on veins; palmately 13-15-veined, veins radiating from petiole tip, and 2 pairs originating from central vein. Inflorescence erect, umbellate, seem to be axillary, but are reduced, sympodial branches with very short internodes, without leaves; common peduncle 2-8 cm long, peduncles slender, 1-1.5 cm long; spikes numerous, 5-10 cm long, each subtended by a single prophyll, densely flowered; floral bracts marginally fringed. Fruits trigonous, glabrous.

Distribution: Widely distributed in tropical S America; ca. 60 collections studied (GU: 25; SU: 13; FG: 25).

Specimens examined: Guyana: Ithaca, Berbice R., Grewal 289 (BRG, U); Rupununi Distr., Kanuku Mts., Jansen-Jacobs *et al.* 2236 (BRG, U). Suriname: Wilhelmina Mts., LBB 16263 (BBS, U); Wilhelmina Mts., S slope Juliana peak, Irwin *et al.* 54650 (NY, U). French Guiana: Inini R., De Granville *et al.* 7303 (CAY, P, U); Saül, Görts-van Rijn *et al.* 67 (CAY, NY, U).

Vernacular names: Guyana: popo sakara (van Andel); cow foot leaf (Reinders 7).

Use: As a fish-poison, for that purpose cultivated (in Guyana, NW Distr., according to Archer).

Note: There has been quite a discussion on the correct name of the taxon. I agree with R. Callejas that the taxon can best be placed in *Piper*, not separated as *Lepianthes* or *Pothomorphe. Piper peltatum* together with *P. marginatum*, *P. umbellatum* L. and several other extra-Guianan taxa form a distinct clade within *Piper* (Jaramillo & Manos, Amer. J. Bot. 88: 712. 2001). They do not form a monophyletic group and should not be separated from *Piper*.

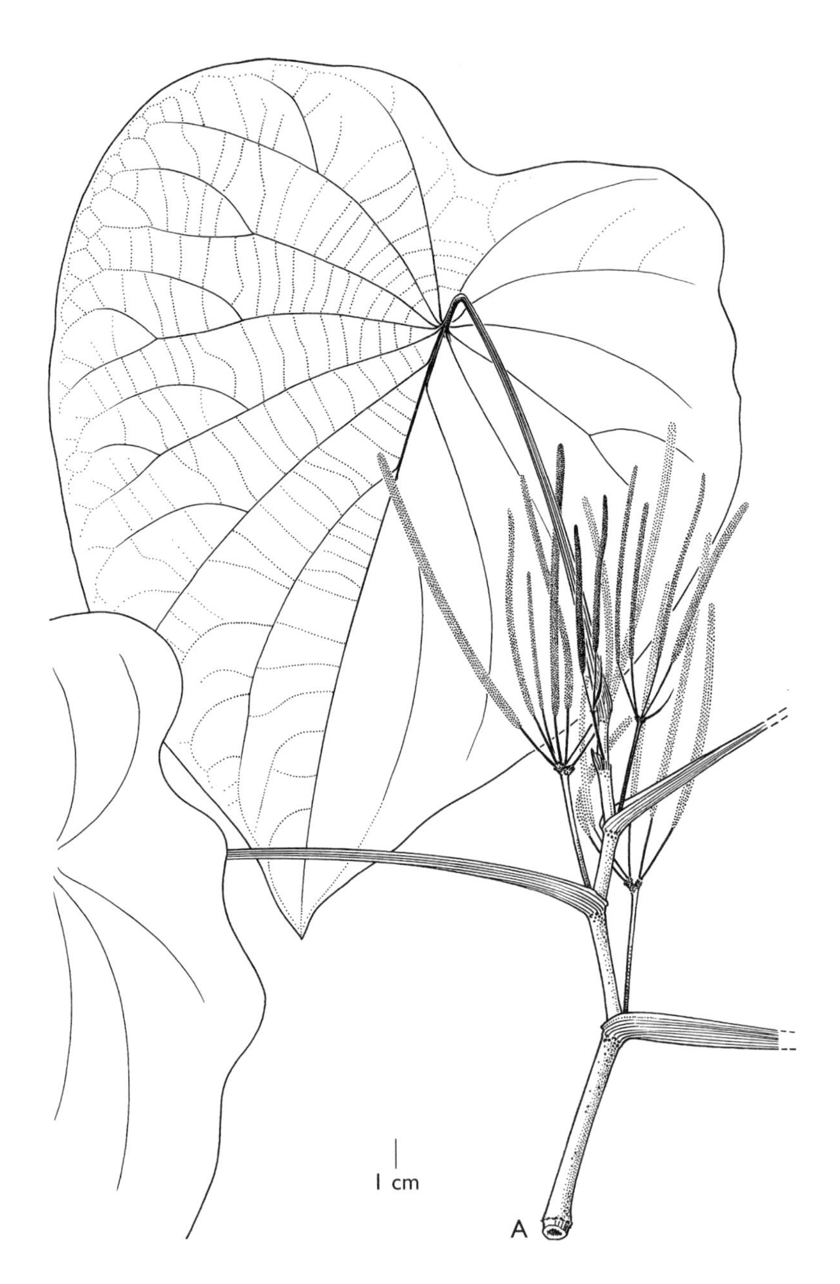

Fig. 50. *Piper peltatum* L.: A, habit (A, Irwin *et al.* 54650).

46. **Piper perstipulare** Steyerm., Fl. Venez. 2(2): 519, f. 78. 1984. Type: Guyana, Upper Mazaruni R., Kako R., camp 3, 0.4 km below the Kako Falls, S.S. & C.L. Tillett 45501 (holotype VEN, not seen, isotype NY). – Fig. 51 A-B

Shrub to 1.5-4 m tall. Stem glabrous. Stipules 4-7 cm long, glabrous, upper ones subpersistent. Petiole (0.5-)1-4.5 cm long, glabrous, vaginate to apex, margin enlarged; blade not scabrous, densely glandular-dotted below, narrowly elliptic to elliptic-ovate, 16-27 x 7-13(-14.5) cm, margin ciliolate, apex acuminate, base equal or unequally attached to petiole difference 0-0.3 cm, obtuse to rounded, glabrous, but veins sparsely to densely appressed or crisp-pubescent below; pinnately veined, secondary veins 6-7 per side, originating from lower $^{2}/_{3}$ of primary vein, flat above, prominulous below, lower 2-3 ascending horizontally or at an angle of 45°, upper ones ascending near margin but hardly anastomosing towards top, tertiary veins reticulate and transverse between secondaries. Inflorescence erect; peduncle 1-2 cm long, glabrous; spike 7-9.5 cm long, greenish white to white, apiculate; floral bracts subtriangular, densely marginally fringed. Infructescence to 11.5 cm long; fruits trigonous or ovoid, glabrous, verruculose, stigmas 3, recurved, sessile.

Distribution: Venezuela, Guyana and French Guiana; from 500-1275 m elev.; 18 collections studied (GU: 16; FG: 2).

Selected specimens: Guyana; Potaro-Siparuni region, Chenapu village, Kvist 284 (U); Pakaraima Mts., Mt. Wokomong, Henkel 4196, 4197 (U); Paruima, Maas *et al.* 5603 (BRG, P, U), Clarke 5433 (U). French Guiana: Saül, Mori *et al.* 22049 (CAY, U), Görts *et al.* 122 (CAY, U).

Note: *Piper perstipulare* can be distinguished from *P. coruscans* and *P. glabrescens,* two other species with rather long-petioled leaves and relatively few secondary veins, by the following combination of characters: It has floral bracts that are densely marginally fringed and leaves that are appressed pubescent on the veins below, whereas *P. glabrescens* has glabrous floral bracts and the leaves below glabrescent. In *P. perstipulare*, the upper secondary veins ascend distinctly to the apex without anastomosing, whereas in *P. coruscans* they anastomose before reaching the apex. A striking difference with *P. coruscans* is that the stipules are longer than the petioles, whereas in *P. coruscans* they are shorter. *Piper perstipulare* has transverse tertiary veins, whereas *P. coruscans* has a less regular pattern.

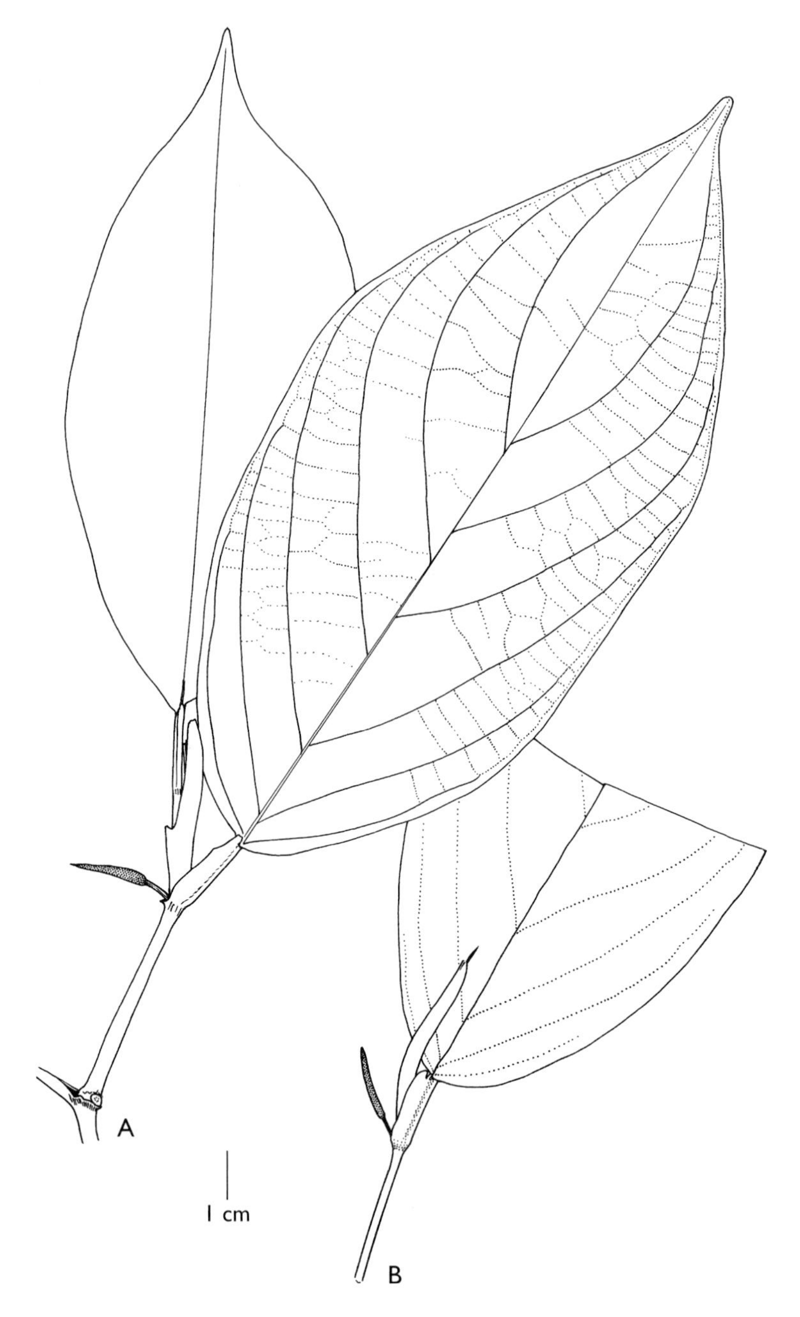

Fig. 51. *Piper perstipulare* Steyerm.: A, habit; B, detail of branch tip (A, Pipoly 10640; B, Pipoly 10786).

47. **Piper phytolaccifolium** Opiz in Presl, Reliq. Haenk. 1: 151. 1828, as 'phytolaccaefolium'. Type: Ecuador, Guayaquil, Haencke 77 (isotype W, not seen). – Fig. 49 B

Shrub to 1.6 m tall, glabrous. Petiole 0.4-1 cm long, vaginate to middle or to apex; blade membranous, not scabrous, glandular-dotted to densely glandular-dotted, elliptic to ovate, 9-16 x 2.5-7 cm, apex acuminate, base unequally attached to petiole difference 0.2-0.3 cm, acute, glabrous; pinnately veined, secondary veins 4-6 per side, originating from lower $^2/_3$ of primary vein, flat above, prominulous below, tertiary veins reticulate. Inflorescence erect; peduncle 0.4-0.7 cm long; spike 2-4 cm long, not apiculate; floral bracts densely marginally fringed. Infructescence erect to pendent, 0.6 cm thick; fruits globose, 1-1.5 mm in diam., glabrous, with persistent style, stigmas 3, on long style.

Distribution: C America, Colombia, Ecuador, Peru, Venezuela and French Guiana; 3 collections studied (FG: 3).

Specimens examined: French Guiana: Saül, de Granville *et al.* 8634 (CAY, NY, U, US).

Note: The persistent styles on the fruit give the plants a striking appearance.

48. **Piper piscatorum** Trel. & Yunck., Piperac. N. South Amer. 395, f. 356. 1950. Type: Venezuela, Bolivár, La Prisión, Medio Cauca, Williams 11685 (holotype US, isotypes F, ILL, not seen). – Fig. 52 A-E

Subshrub, to 1 m tall. Stem glabrous or glabrescent; upper internodes finely striate. Petiole 0.5-2 cm long, vaginate at base; blade subcoriaceous, dull green, not conspicuously glandular-dotted, elliptic-oblong to narrowly ovate, (5.5-)7.5-17(-19.5) x (2.5-)3.5-6.5(-7.5) cm, apex acuminate, base equal or almost equally attached to petiole, subacute to obtuse or rounded, glabrous; pinnately veined, secondary veins (5-)7-8(-11) per side, originating from throughout primary vein, anastomosing well within margin, impressed or slightly prominulous above, prominent below, tertiary venation obsolete. Inflorescence erect; peduncle 0.4-0.5(-1) cm long, glabrous; spike 4-7 cm long, green, apiculate; rachis pubescent; floral bracts cucullate, glabrous. Infructescence to 6 cm long; fruits separate at maturity, not sulcate, globose, 1.5 mm in diam., papillose at apex, glabrescent, stigmas 4, sessile.

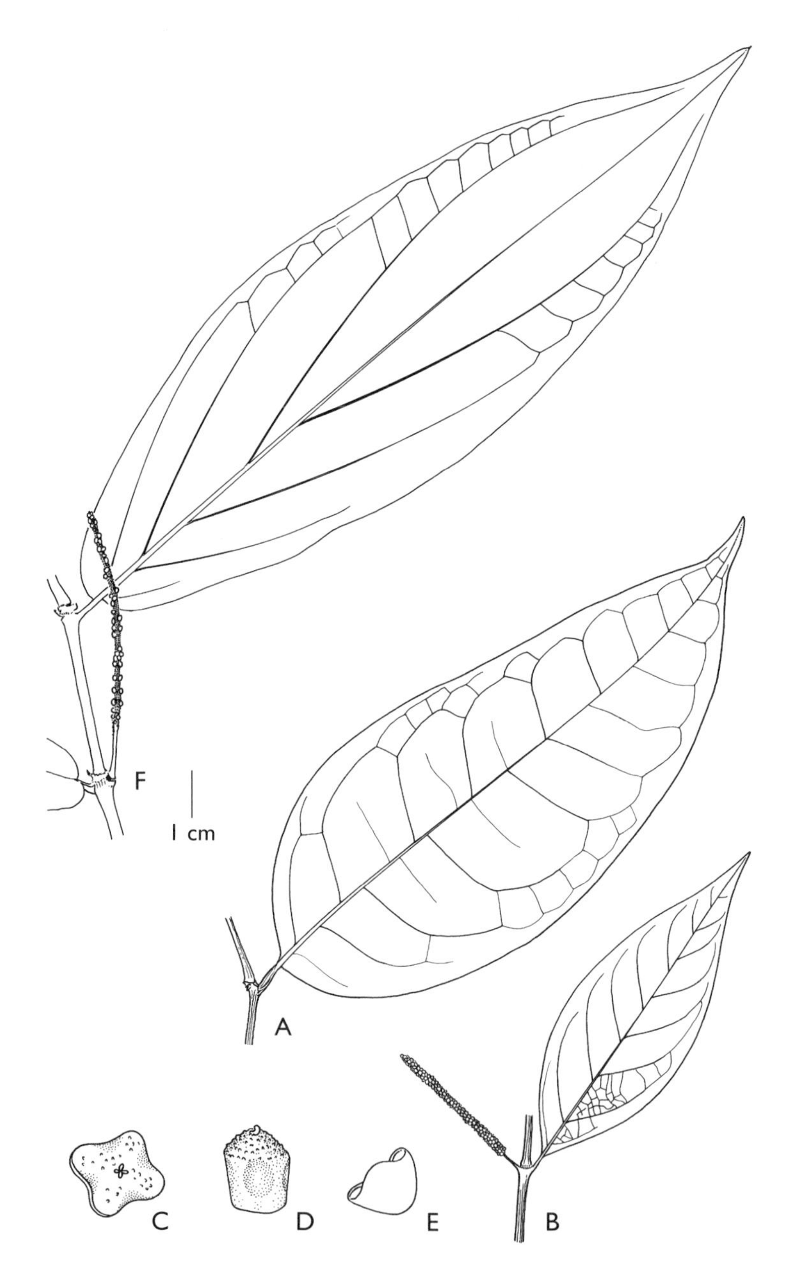

Fig. 52. *Piper piscatorum* Trel. & Yunck.: A, leaf; B, part of branch with leaf and spike; C, fruit (seen from above); D, fruit (lateral view); E, floral bract. *Piper pulleanum* Yunck.: F, part of branch with leaf and spike (A, Williams 11685; B-E, based on fig. 79 in Steyermark, 1984; F, Clarke 4539).

D i s t r i b u t i on: Venezuela and Brazil (Mato Grosso); non-flooded moist forest, or along riverbanks, up to 660(-1100?) m elev.; not yet reported from the Guianas, but expected to be used there too as fish poison.

Uses: Roots and branches are used as fish poison; also used to alleviate pain of teeth and gums. (McFerren, M.A. & E. Rodriguez, J. Ethnopharmacology 60: 183-187. 1998).
According to R. Callejas, the plants have an anaesthetic effect.

Note: For differences with *Piper aulacospermum*, see note to the latter.

49. **Piper poiteanum** Steud., Nomencl. Bot. ed. 2. 2: 342. 1841. – *Enckea dubia* Kunth, Linnaea 13: 606. 1840 ('1839') (non *Piper dubium* A. Dietr. 1831). Type: French Guiana, Poiteau s.n. (holotype B, not seen, isotypes K, U). – Fig. 49 C

Vine or climbing shrub to 8 m tall. Upper internodes finely striate, glabrescent. Petiole 1.5-2.5 cm long, glabrous or sparsely pubescent, vaginate to apex; blade membranous, bicolourous, not scabrous, not conspicuously glandular-dotted, oblong to ovate, 7-12 x 3-5 cm, apex acuminate, base equal, rounded to subcordate to deeply cordate, glabrous but veins glabrous or sparsely pubescent below; palmately-pinnately veined, secondary veins 2-3 per side originating from lower $1/4$ of primary vein, flat above, prominent below, tertiary veins transverse, prominulous below. Inflorescence presumably pendent; peduncle 1-1.5 cm long, minutely pubescent or glabrous; spike 7-16.5 cm long, yellow to green, apiculate; floral bracts cucullate-inflexed, densely marginally fringed. Infructescence to 15 cm long; fruits depressed globose to trigonous, 1 mm thick, glabrous, stigmas 3, sessile.

Distribution: The Guianas and Venezuela (Amazonas, Bolivar); in understorey of primary rainforest and from secondary vegetation, from 10-600 m elev.; 12 collections studied, several sterile (GU: 5; SU: 5; FG: 4).

Fertile specimens examined: Guyana: Berbice-Courantyne region, near Cow falls, McDowell 2227 (BRG, U, US); Kanuku Mts., Crabwood Cr., Jansen-Jacobs *et al.* 4363 (BRG, U). Suriname: locality unknown, Kappler 1482 (P, U); Brownsberg Nature Reserve, Lindeman & Heyde 695 (BBS, U). French Guiana: Piste de St Elie, Prévost 1648 (CAY, U).

Note: Steyermark & Callejas (2002: 719) erroneously used *Piper foveolatum* C. DC. as name for this taxon, apparently not realizing that Steudel in 1841 already validly published a combination in *Piper*.

50. **Piper pulleanum** Yunck. in Trel. & Yunck., Piperac. N. South Amer. 323. 1950. Type: Suriname, Upper Saramacca R., Pulle 446 (holotype U). – Fig. 52 F

Shrub or treelet, to 2.5 m tall. Stem densely retrorsely crisp-pubescent. Stipules may be long persistent. Petiole 0.3-0.7(-1) cm long, densely brown-pubescent; blade not conspicuously glandular-dotted, often drying dull greyish green, somewhat scabrous and bullate, ovate or elliptic-ovate, 11-25 x 4.5-8 cm, apex acuminate, base unequally attached to petiole difference 0.1-0.3 cm, obtuse to subcordate, almost glabrous above, veins may be somewhat appressed-pubescent, appressed-pubescent below especially on veins; pinnately veined, secondary veins 3-4(-5), originating from lower 1/2 of primary vein, ascending at an angle of ca. 45°, flat to prominulous above, prominent below, tertiary veins transverse. Inflorescence erect; peduncle 0.5-8(-2.5) cm long, brown crisp-pubescent; spike 5-8 cm long, white to yellowish, apiculate; floral bracts marginally fringed; anthers dehiscing laterally. Infructescence erect; fruits obovoid to trigonous, papillose or hispidulous, ca. 2 mm wide; stigmas 3.

Distribution: The Guianas; to 260 m elev.; 22 collections studied (GU: 7; SU: 5; FG: 10).

Selected specimens: Guyana: Rupununi Distr., Kuyuwini R, Kuyuwini Landing, Jansen-Jacobs *et al.* 3152 (BBS, BRG, P, U), Kuyuwini R., Clarke 4516 (U, US). Suriname: Litanie R., near Feti Cr., Geijskes 81 (U); Oelemari R., near airstrip, Wessels Boer 965 (U). French Guiana: Mt. St Michel, de Granville *et al.* T-1164 (CAY, P, U); Upper Marouini R. basin, Oeman fou langa soela, de Granville *et al.* 9202 (B, CAY, INPA, NY, P, U, US).

Vernacular name: petpe (Wayana). This name is not unique for the taxon.

51. **Piper remotinervium** Görts, Blumea 50: 372. 2005. Type: French Guiana, Saül, route de Bélizon, pk 7, Eau Claire, de Granville *et al.* 4474 (holotype U, isotypes CAY, P). – Fig. 53 A-H

Shrub or treelet, to 1-3.5 m tall. Stem and branches sparsely to densely pubescent. Petiole 1.5-2 cm long, densely pubescent; blade somewhat scabrous when dried, not conspicuously glandular-dotted, broadly ovate or ovate, 17-30 x10-17 cm, apex acuminate, base equal or unequally attached to petiole difference 0-0.3(-0.7) cm, subcordate or obtusish, pubescent on both surfaces, veins more densely so; pinnately veined,

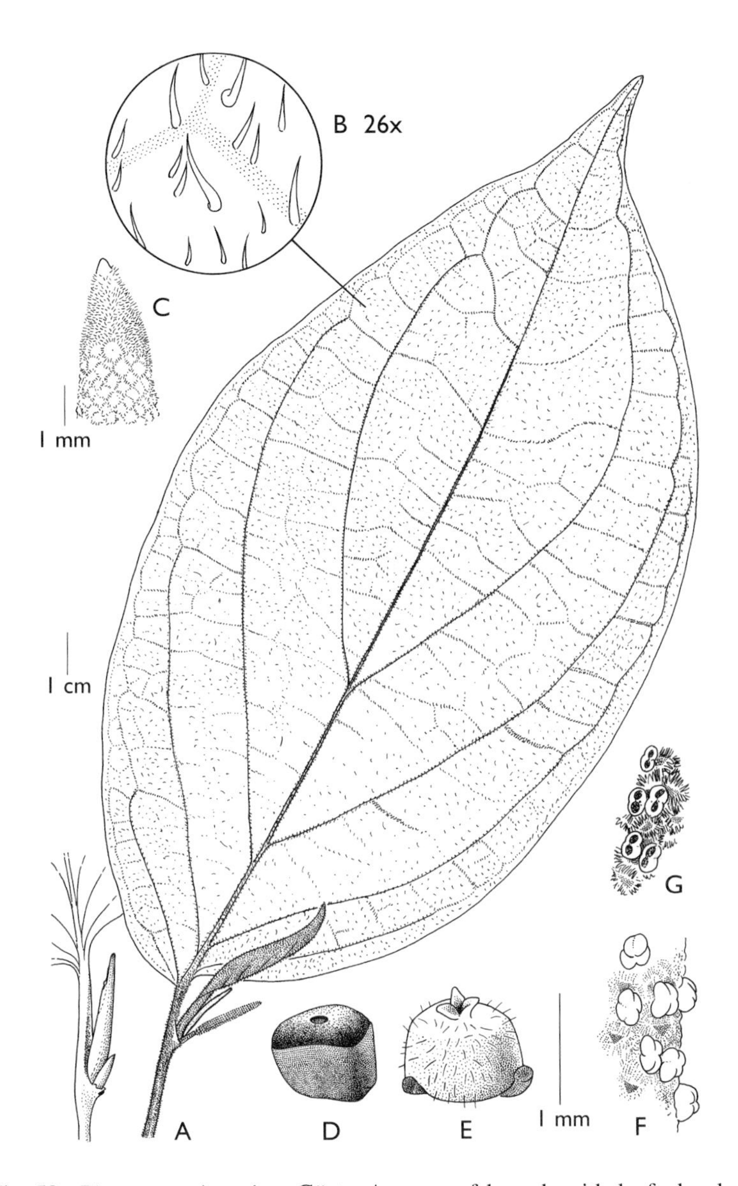

Fig. 53. *Piper remotinervium* Görts: A, apex of branch with leaf, developing young branch tip, prophyll and immature spike; B, indument on lower leaf surface; C, apiculate tip of spike; D, fruit as seen in dried collection; E, fruit with persistent stigma's and remnants of stamens; F, detail of spike showing young anthers; G, detail of spike with open anthers between fimbriate floral bracts; H, apex of young branch showing leaf base, developing tip of branch and prophyll (A-C, Feuillet 530; D, de Granville *et al.* 4444; E, de Granville *et al.* 13110; F-G, McDowell 2225; H, Cremers 15083).

secondary veins 5-7 per side, originating from lower 1/2 of primary vein, prominent below, tertiary venation widely reticulate, veinlets almost transverse. Inflorescence erect or horizontally oriented; peduncle 1.5-2 cm long, pubescent; spike 10-20 cm long, creamish white to green, hardly apiculate; floral bracts densely marginally fringed. Fruits trigonous, somewhat exserted when mature, hirsute, stigmas sessile.

Distribution: French Guiana and Brazil (Pará); in (dense) forest, 20-300 m elev.; 20 collections studied (FG: 34).

Selected specimens: French Guiana: Ile de Cayenne, de Granville *et al.* 9128 (CAY, NY, P, U, US); Kaw Mts., Feuillet 2233 (CAY, NY, U, US), Hoff 5540 CAY, U, US); Saül, de Granville *et al.* 9052 (CAY, MO, NY, P, U, US), Mori *et al.* 21994 (CAY, NY, U).

Note: The specimens here assigned to *Piper remotinervium* previously had erroneously been identified as *P. tectoniifolium* (Kunth) Steud. The latter species, however, is endemic to the Atlantic forests of Brazil. It grows in open non-flooded areas or forest margins.

52. **Piper reticulatum** L., Sp. Pl. 29. 1753. Type: [icon] Plumier, Descr. Pl. Amér. 7: t. 75. 1693 (according to Howard 1973: 407).
– Fig. 54 A

Treelet or shrub, sometimes scandent, to 10 m, tall. Stem often with spines opposite petiole; nodes glandular-dotted. Prophyll oblong, 0.4-1 cm long. Petiole 3-12 cm long, glabrous, vaginate at base; blade membranous or firmly chartaceous, not glandular-dotted, ovate or broadly ovate, 14-30 x 7-18 cm, apex acute or acuminate, base equal, obtuse or rounded; palmately 5-7(-9)-veined, veins plane to impressed above, very prominent below, tertiary veins transverse. Inflorescence erect; peduncle 0.7-2 cm long, glabrous or minutely puberulous; spike 6-12(-15) cm long, white or greyish, not aciculate; rachis glabrous; floral bracts glabrous or sparsely puberulous. Fruits greenish white, obovoid or oblongoid, 1.5 x 1.2-1.5 mm, puberulous, with glabrous disc at apex, separate at maturity, stigmas 3(-4), sessile.

Distribution: West Indies, C and northern S America; recorded mainly from secondary vegetation, to 450 m elev.; 18 collections studied (GU: 3; SU: 3; FG: 9).

Selected specimens: Guyana: Kanuku Mts., Maipaima, Jansen-Jacobs *et al.* 1059 (BBS, BRG, P, U); Marudi Mts., Mazoa Hill, Stoffers & Görts-van Rijn *et al.* 227 (BBS, BRG, U). Suriname: Nickerie Distr.,

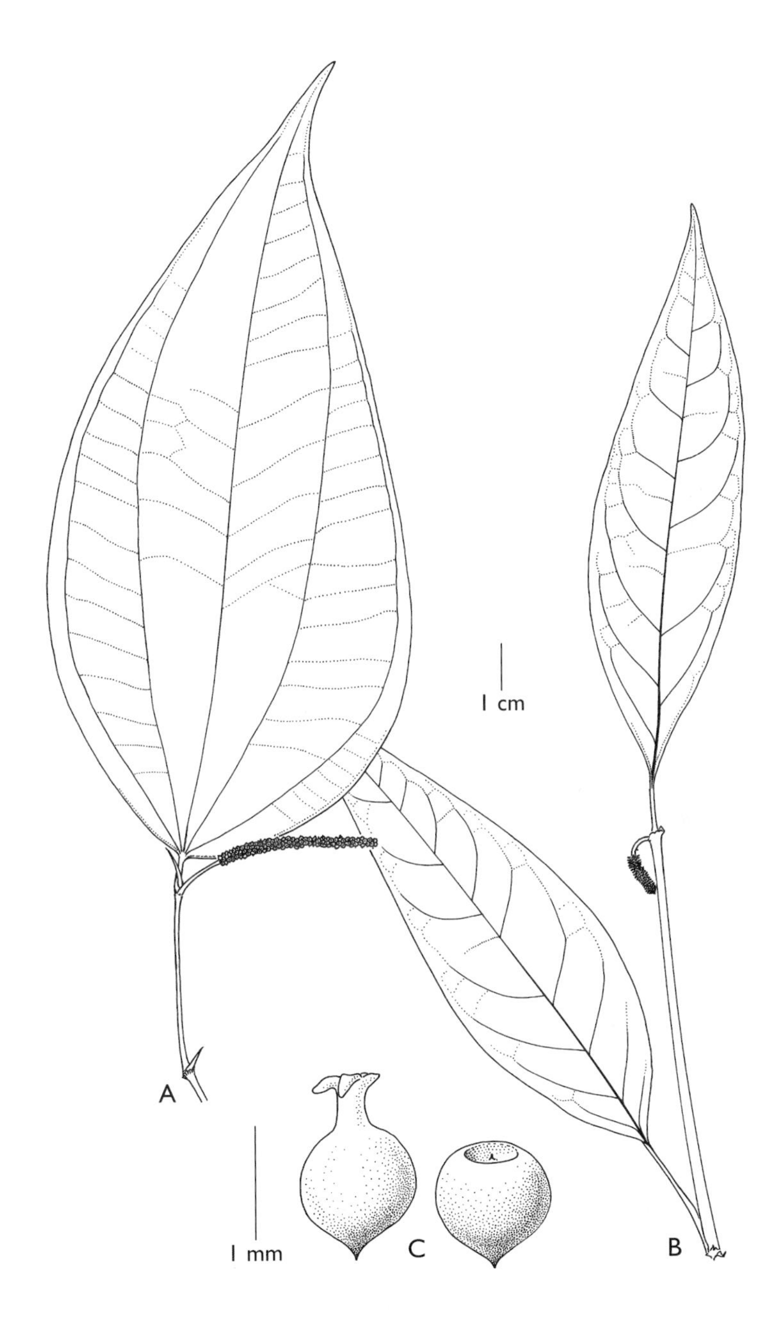

Fig. 54. *Piper reticulatum* L.: A, habit. *Piper rudgeanum* (Miq.) C. DC.: B, habit; C, fruits (A, de Granville *et al.* 2288; B-C, Kegel s.n. type).

Kabalebo Dam area, Lindeman & de Roon *et al.* 801 (BBS, U), Lindeman & Görts-van Rijn *et al.* 640 (BBS, U). French Guiana: Saül, de Granville *et al.* 2288 (CAY, P, U), Philippe *et al.* 27011 (ILL, NY, U).

Vernacular name: aneisi wiwiri (Sranan). This name is not unique for the taxon.

53. **Piper rudgeanum** (Miq.) C. DC. in A. DC., Prodr. 16(1): 373. 1869. – *Artanthe rudgeana* Miq., Linnaea 22: 77. 1849. Type: Suriname, near Loango, Kegel s.n., Ao june 1846 (holotype U). – Fig. 54 B-C

Shrub. Stem glabrous, glandular. Petiole 0.3-2 cm long, vaginate to apex; blade not scabrous, densely glandular-dotted on both surfaces, narrowly ovate or lance-elliptic, 9-13 x 2-4 cm, apex acuminate, base almost equal, acute to cuneate, glabrous; pinnately veined, secondary veins 7-10 per side, originating from throughout primary vein, flat above, flat below, marginally loop-connected, tertiary venation reticulate. Inflorescence pendent; peduncle rather stout, 0.5-0.7 cm long, glabrous; spike 1 cm long, not apiculate; floral bracts pilose on inner side and at base. Infructescence recurved, 1 cm long, 0.3 cm thick; fruits ovoid, 2 mm in diam., glabrous, with persistent short style, stigmas 3-4 recurved on short, persistent style.

Distribution: Only known from the type collection from Suriname.

Notes: Yuncker mentioned the type collection in the Flora of Suriname. No further collections have been reported since.
Piper rudgeanum resembles *P. nigrispicum* in most vegetative characters, although the type shows more secondary veins. See also note to *P. nigrispicum*.

54. **Piper rupununianum** Trel. & Yunck., Piperac. N. South Amer. 303, f. 268. 1950. Type: Guyana, Tumatumari, Gleason 107 (holotype NY, isotypes K, US). – Fig. 55 A

Shrub or tree, 1-2 m tall. Stem glabrous or may be sparsely appressed-pubescent. Petiole 0.3-0.6 cm long, glabrous, vaginate usually at base; blade not glandular-dotted, narrowly elliptic or narrowly elliptic-oblong, 14-20 x 5-7 cm, apex long-acuminate, base equal or almost equally attached to petiole, acute, glabrous; pinnately veined, secondary veins 3-4 per side, originating from lower $1/2$ to lower $2/3$ of primary vein, ascending at an angle of 45-60°, plane above, prominulous below,

Fig. 55. *Piper rupununianum* Trel. & Yunck.: A, habit (A, Henkel 4200).

tertiary veins widely reticulate. Inflorescence erect or pendent; peduncle 1-1.5(-3) cm long; spike 5-9 cm long, apiculate; floral bracts densely marginally fringed. Fruits trigonous to obpyramidal, glabrous, glandular, green, stigmas 3, minute, sessile.

Distribution: Venezuela, Guyana and French Guiana; in clearings, damp places, open and mixed forest, shady rocky areas, from sea level to 1600 m elev.; 31 collections studied (GU: 25; FG: 6).

Selected specimens: Guyana: Potaro-Siparuni region, Mt. Kopinang, Hahn 4300 (U, US); Cuyuni R., Tutin 318 (BM); Bartica, Bartlett 8408 (BRG, K, U). French Guiana: Trinité Mts., de Granville *et al.* 13225 (U); without locality, herb. Sagot s.n. (P).

55. **Piper salicifolium** Vahl, Enum. Pl. 1: 312. 1804. – *Artanthe salicifolia* (Vahl) Miq., Syst. Piperac. 533. 1844. Type: Suriname, Rolander s.n. (holotype C). – Fig. 56 A-B

Shrub. Stem retrorsely villous. Petiole ca. 0.5 cm long, villous; blade somewhat scabrous above, conspicuously dark-glandular-dotted below, narrowly elliptic or narrowly ovate, 5-7.5 x 1-1.5 cm, margin ciliate, apex long-acuminate, base equal, obtuse or subacute, densely villous below; pinnately veined, secondary veins ca. 5 per side, originating from lower $^{1}/_{2}$ of primary vein, tertiary veins more or less parallel, prominulous below. Inflorescence: peduncle ca. 1 cm long; spike ca. 4 cm long, 0.3 cm thick; floral bracts densely marginally fringed; ovary and fruit too young to describe accurately.

Distribution: Known only from the type collection from Suriname.

Note: There is still no recent material that could be identified as *Piper salicifolium.* It differs from the other narrow-leaved species in having somewhat scabrous leaf blades that are densely villous below. The leaves has few secondary veins that originate from the lower half of the primary vein.

56. **Piper trichoneuron** (Miq.) C. DC. in A. DC., Prodr. 16(1): 281. 1869. – *Artanthe trichoneura* Miq., Syst. Piperac. 483. 1844. Type: French Guiana, Leprieur 143 (holotype G-DEL, isotype NY not seen, P) . – Fig. 57 A-C

Artanthe berbicensis Miq., Syst. Piperac. 500. 1844. – *Piper hostmannianum* (Miq.) C. DC. var. *berbicense* (Miq.) C. DC. in A. DC., Prodr. 16(1): 287. 1869. Type: Guyana, Schomburgk s.n. (holotype B, not extant).

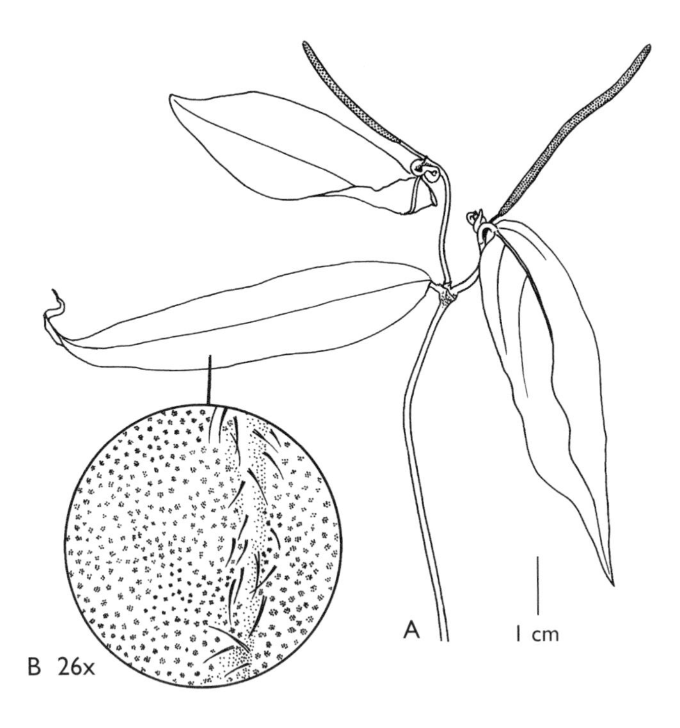

Fig. 56. *Piper salicifolium* Vahl: A, habit; B, detail of leaf lower surface with indument and glandular dots (A-B, Rolander s.n., holotype).

Artanthe kegeliana Miq., Linnaea 22: 77. 1849. – *Piper kegelianum* (Miq.) C. DC. in A. DC., Prodr. 16(1): 372. 1869. Type: Suriname, Para, near Klein Onoribo, Kegel 659 (holotype U, isotype G), syn. nov.
Piper gleasonii Yunck. in Maguire *et al.*, Bull. Torrey Bot. Club 75: 287. 1948. Type: Guyana, Rockstone, Gleason 857 (holotype US, isotypes GH, K, NY, U), syn. nov.

Shrub, sometimes treelet, to 2 m tall. Stem crisp-pubescent. Stipules late deciduous. Petiole often stout, to 1 cm long, densely crisp-pubescent, vaginate to apex; blade sometimes with pale band along primary vein, coriaceous, more or less orange glandular-dotted on lower surface, sometimes somewhat scabrous, (narrowly) elliptic, elliptic-oblong or (narrowly) ovate, 12-33 x 5-13 cm, apex (long-)acuminate, base equal or unequally attached to petiole difference 0-0.5 cm, obtuse to rounded to subcordate occasionally cuneate, glabrous above, crisp-pubescent on veins below; pinnately veined, secondary veins 4-7(-11) per side, originating from lower $^{1}/_{2}$ to $^{2}/_{3}$ of primary vein, not anastomosing, impressed above, rather prominent below, tertiary veins widely

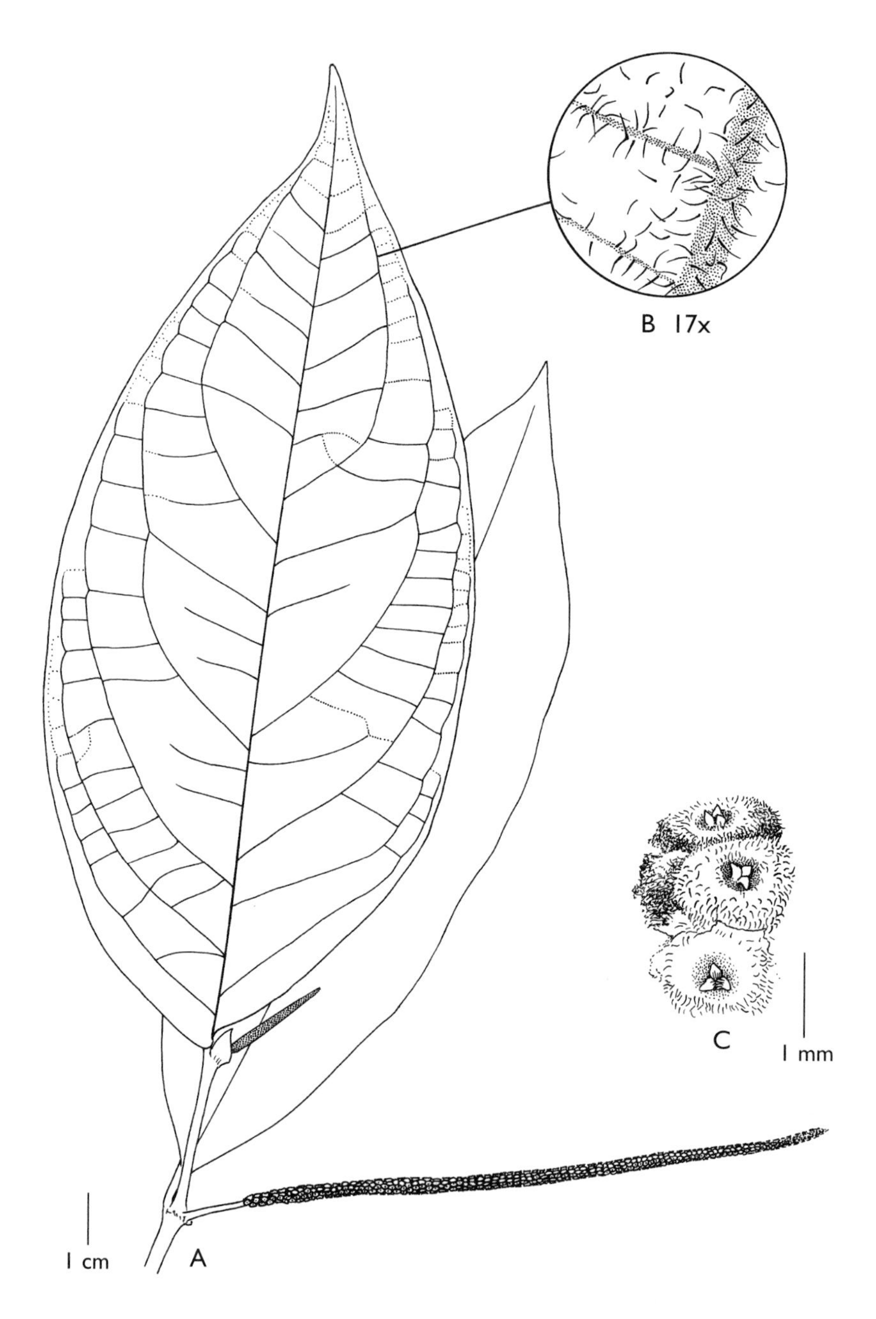

Fig. 57. *Piper trichoneuron* (Miq.) C. DC.: A, habit; B, indument on lower leaf surface; C, fruits (A-C, Lindeman 562).

reticulate, transverse. Inflorescence erect; peduncle 1 cm long to ca. 2 cm in fruit, pubescent; spike 8-9 cm long, in fruit to 11 cm long, often reddish or pink, apiculate; floral bracts densely marginally fringed. Fruits oblongoid, glandular, hirsute, stigmas 3, obtuse, sessile.

Distribution: The Guianas and Brazil (Acre); over 110 collections studied (GU: 29; SU: 28; FG: 56).

Selected specimens: Guyana: Rupununi Distr., between Kuyuwini Landing and Kassikaityu R., Jansen-Jacobs *et al.* 3034 (BRG, U); Rupununi Distr., Shea to Quitaro R., Bowi Cr., Jansen-Jacobs *et al.* 4891 (BRG, U). Suriname: Wilhelmina Mts., Irwin *et al.* 54611 (K, U); Kabalebo dam area, Lindeman & Görts-van Rijn *et al.* 246 (BBS, U). French Guiana: Saül area, Görts-van Rijn *et al.* 55 (CAY, NY, U, US); Mt. Bellevue de l'Inini, de Granville *et al.* 7946 (B, CAY, INPA, NY, MG, MO, U).

Notes: The leaf base of *Piper trichoneuron* is very variable in shape. *Piper trichoneuron* and *P. hostmannianum* resemble each other in having pubescent, coriaceous leaves. See also note to *P. hostmannianum*.
Tebbs (1993: 44) includes *P. gleasonii* in *P. jacquemontianum*, a C American species. In my opinion *P. gleasonii* is similar to *P. trichoneuron* and thus I include it under the latter.

57. **Piper tuberculatum** Jacq., Icon. Pl. Rar. 2: 2, t. 211. 1795. – *Piper arboreum* Aubl. subsp. *tuberculatum* (Jacq.) Tebbs, Bull. Brit. Mus. (Nat. Hist.), Bot. 19: 156. 1989. Type: Venezuela, Caracas, Bredemeyer s.n. (holotype W).

Shrub or treelet to 3 m tall, stem minutely hirtellous, usually somewhat tuberculate. Petiole tuberculate, 0.1-0.7 cm long, minutely puberulent, glabrescent, vaginate to base of blade and extending shortly; blade broadly ovate-elliptic, 4-12 x 2-6 cm, apex obtuse, rounded or slightly acute, base unequally attached to petiole difference 0.2-0.7 cm, shorter side acute to obtuse, longer side rounded or cordate, glabrous or glabrescent above, minutely pubescent on veins below; pinnately veined, secondary veins 4-10, originating from throughout primary vein, anastomosing near margin. Inflorescence erect, 4-14 cm long; peduncle 0.5-2 cm long; spike 2-11 cm long; floral bracts 0.5-0.7 mm in diam, marginally fringed, conspicuously arranged in whorls. Fruits laterally compressed, rounded, glabrous, stigmas 3-4, sessile.

Distribution: S Mexico, C America and northern S America; mostly in dry areas; in the Guianas reported only from S Guyana and Suriname; 10 collections studied (GU: 6; SU: 3).

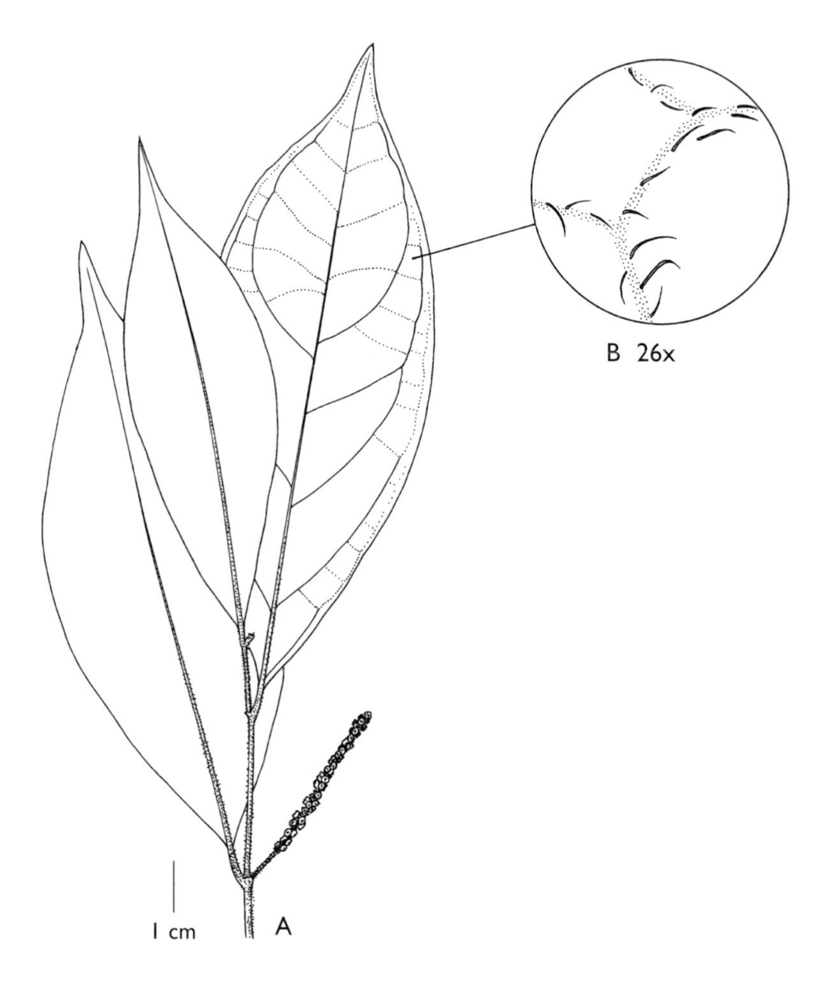

Fig. 58. *Piper wachenheimii* Trel.: A, habit; B, indument on lower leaf surface (A-B, de Granville *et al.* 7229).

Specimens examined: Guyana: Kanuku Mts., A.C. Smith 2529 (P, U), 3366 (P, U), 3436 (NY, P, U); Moco moco R., Jansen-Jacobs *et al.* 4442 (BBS, BRG, U). Suriname: near Paramaribo, Kramer & Hekking 2043 (BBS, U); Suriname R., Plantation La Liberté, Soeprato 226 (U); Marataka R., near Cupido, BW 841 (BBS, U).

Note: *Piper tuberculatum* and *P. arboreum* share the same striking character of a very unequal leaf base. Difference at the petiole up to 3 cm. They can, however, be distinguished by the shape and presence of indument on the leaves. In *P. arboreum* internodes and petioles are smooth or with a few tubercles only; blades are ovate to elliptic, the apex acute to (long-) acuminate, whereas in *P. tuberculatum* petioles are usually tuberculate, blades relatively smaller and more narrowly ovate-elliptic and the leaf apex obtuse, rounded or only slightly acute. Like several authors (Steyermark 1984: 583; Trelease & Yuncker 1950: 364, f. 330; Burger 1971: 185, f. 8 and Steyermark & Callejas 2003: 726) I consider *P. tuberculatum* to be a separate species. This in contrast to Tebbs, who found too few differences to keep *P. tuberculatum* and *P. arboreum* apart; as they occur in different habitats, Tebbs accepted the two species as subspecies. *Piper tuberculatum* occurs in dry habitats in contrast to *P. arboreum* which prefers more moist habitats, swamp or mixed forest and secondary vegetation.

58. **Piper wachenheimii** Trel., Bull. Misc. Inform. Kew 1933: 339. 1933. Type: Guyana, near Bartica, Sandwith 231 (holotype K, isotype NY). – Fig. 58 A-B

Small shrub or even small treelet, to 1.5 m tall, nodose in lower part. Stem densely pubescent or hirsute, hairs often reddish brown (to 0.7 mm long). Petiole 0.3-0.5 cm long, densely pubescent or hirsute, vaginate at base; blade not scabrous, glandular-dotted, lance-elliptic to oblanceolate, 12-15 x 3.5-5 cm, apex acuminate, base equal or almost equally attached to petiole, acute or cuneate, glabrous above, erect-pubescent below, more so on veins; pinnately veined, secondary veins 4-5 per side, originating from lower $^2/_3$ of primary vein, not anastomosing, plane to impressed above, prominulous or prominent below, tertiary veins reticulate. Inflorescence erect; peduncle to 0.5 cm long, crisp-pubescent; spike to 8(-12?) cm long, reddish, pinkish or brownish, apiculate; rachis glabrous; floral bracts densely marginally fringed. Fruits depressed globose or obovoid, ca. 1 mm thick, pubescent or hirtellous, separate from each other, stigmas 3, sessile.

Distribution: Brazil (Pará) and the Guianas; in primary forest, and occasionally from secondary vegetation, from sea level to 600 m elev.; ca. 58 collections studied (GU: 13; SU: 10; FG: 35).

Selected specimens: Guyana: NW Distr., Kwabanna, Waini Cr., Maas *et al.* 2427 (BRG, NY, U); Mazaruni Station, Forest Dept. 2516 (U). Suriname: Ulemari, UVS 17815 (BBS, U); Brokopondo, van Donselaar 2468 (U). French Guiana: Ouaqui R., de Granville *et al.* 1764 (CAY, P, U); Petit saut de Sinnamary R., Prévost 1419 (CAY, P, U).

Vernacular name: Guyana: warabakakoro (Arawak). This name is not unique for the taxon.

U s e : The plants are used in religious ceremonies by Suriname Hindustani.

N o t e s : Recognisable by the symmetrical, acute leaf base and the erect pubescence on the lower leaf surface and veins. The plants have a strong peppery smell (according to T. van Andel, pers. comm.).
The collection French Guiana, Wachenheim 168 in Paris bears a type label. Although Trelease mentioned this collection together with Sandwith 231 in his protologue, he designated the latter as the type. The Paris collection thus is a paratype.
The Sagot 844 specimens in NY and P belong to *Piper wachenheimii*, this in contrast to the B specimen on which C. de Candolle based his *P. rubescens*; see also note to *P. demeraranum*.

DOUBTFUL SPECIES

Peperomia schomburgkii C. DC. in A. DC., Prodr. 16(1): 395. 1869 was based on the Guyanan collection Ri. Schomburgk 406 seen in the Berlin herbarium. *P. schomburgkii* was described as a glabrous herb, having long petiolate, suborbicular leaves with emarginulate apex, glabrous on both sides; the ovary ending in a short style with apical stigma. No further collections have been reported since. More collections are needed to circumscribe and differentiate the taxon. It may be a distinct taxon, but it may also be an atypical specimen belonging to *P. quadrifolia*.

Piper angremondii C. DC. in Schinz, Vierteljahrsschr. Naturf. Ges. Zürich 60: 431. 1915. Type: Suriname, Suriname R., Plantage Accaribo, d'Angremont s.n. (holotype Z, not seen). Without having seen the holotype, no final conclusion on this species can be drawn. In the literature, it is usually listed as a synonym of *P. hispidum*, whereas on the website of Z, *P. dilatatum* is given as the accepted name. These two species have a superficial resemblance.

Piper laevigatum Kunth in Humb., Bonpl. & Kunth, Nov. Gen. Sp. ed. qu. 1: 56. 1816.
Lemée (1956: 27) mentioned a specimen of *P. laevigatum* Kunth from Cayenne as present in the Herbarium of the Museum (= P, I suppose). I have not seen this specimen, and taking into account the distribution of *P. laevigatum* (Colombia, Eduador, Peru and Bolivia), its occurence in French Guiana is rather unprobable.

Piper microstachyon Vahl, Eclog. Amer. 2: 3. 1798 has never been reported after it was described from an incomplete specimen. I agree with Trelease and Yuncker (1950: 418) that this collection by von Rohr s.n. from Cayenne, French Guiana may well be assigned to *P. angustifolium*.

Piper modestum (Miq.) C. DC. in A. DC., Prodr. 16(1): 275. 1869. – *Artanthe modesta* Miq., Syst. Piperac. 517. 1844. Type: French Guiana, Leprieur s.n. (G-DEL). This is an insufficiently known taxon. The holotype was not seen, only a fragmentary isotype from which no conclusion can be drawn.

Piper purpurascens Desf., Cat. Pl. Hort. Reg. Paris. 414. 1829 was mentioned by Lemée (1956: 26) for French Guiana. I have not seen any specimen and I have no idea which species is concerned. C. DC. (1869: 380) already placed it in the category Species dubiae.

P. schlechtendahlianum C. DC. in A. DC., Prodr. 16(1): 324. 1869 is a nom. illeg. based on among others *Enckea schlechtendalii* Miq. (which was based on specimens from Mexico) and "*P. angustifolium* Pav." (an invalid name, probably a nomen herbariorum). I have not seen the type of Miquel's species name, and it is not clear which species Lemée had in mind when he (1955: 481) accepted this species for French Guiana: "*P. angustifolium* Pav. (P. *schlechtendalianum* or *schlechtendalii* Pav.)".

WOOD AND TIMBER

HERNANDIACEAE

by

ALBERTA M.W. MENNEGA[4]

WOOD ANATOMY

The wood of the two genera indigenous in the Guianas, *Hernandia* and *Sparattanthelium*, is very much alike, as well in general aspects as in structure. The bark is thin and superficially longitudinally grooved; the beige wood is rather coarse, particularly in *Hernandia*, and light. Special features are absent.
The wood is of no commercial value.

FAMILY DESCRIPTION

Growth rings present and rather distinct, boundaries indicated by a few rows of flattened fibres and by an almost continuous ring of aliform/confluent parenchyma.
Vessels diffuse, for 50-80% solitary, the remainder in radial pairs occasionally in clusters or in longer chains, scarce, average number in *Hernandia* 1.5 per mm^2, in *Sparattanthelium* 5 per mm^2. Perforations simple. Intervessel pits alternate, 10-14 µm wide, the apertures included. Vessel member length in *Hernandia* 580 µm, in *Sparattanthelium* 370 µm. Vessel-ray pits very large, oval to irregularly shaped, in *Sparattanthelium* more or less in scalariform arrangement, in *Hernandia* also partly similar to the intervessel pitting.
Rays 1-4-seriate, uniseriates rather scarce; 3-4 per mm; almost homocellular, marginal cells of the same height as procumbent cells, but radially slightly shorter; width 20-50 µm, height up to 600 µm (18 cells) in *Hernandia*, up to 235 µm (30 cells) in *Sparattanthelium*.
Parenchyma rather abundant, mainly paratracheal, vasicentric to aliform and aliform-confluent, also apotracheal diffuse and diffuse-in-aggregates; strands of 2-4 cells. Large cystoliths in inflated cells in *Sparattanthelium p.p.*
Fibres non-septate, thin-walled with a wide lumen, diam. 20-25 µm. Pits simple, minute, in radial walls. Average length of fibres in *Hernandia* 486 µm, in *Sparattanthelium* 860 µm.

[4] Nationaal Herbarium Nederland, Utrecht University branch, Heidelberglaan 2, 3584 CS Utrecht, The Netherlands

GENERIC DESCRIPTIONS

HERNANDIA L. Fig. 59 A

Growth rings usually present, boundaries indicated by a few to several rows of flattened fibres, occasionally by a more or less continuous band of parenchyma.
Vessels diffuse, for the greater part solitary (50-80%), the remainder in radial pairs, seldom in longer chains or in clusters, outline round to oval, 160-250 µm wide; number very scarce, 1.5 (0-3) per mm^2. Vessel member length 580 (300-840) µm. Intervessel pits large, alternate, 14-17 µm with included, wide apertures. Vessel-ray pits and vessel-parenchyma pits of two types, similar to the intervessl pits and very large and irregular-shaped, with strongly reduced borders.
Rays 1-4-seriate, uniseriates rather scarce; 4 (2-5) per mm; almost homocellular, marginal cells of the same height as the procumbent cells; multiseriates usually with short uniseriate margins; height up to 600 µm (18 cells).
Parenchyma mainly paratracheal, ranging from vasicentric to aliform and aliform-confluent, also some diffuse and in aggregates; strands of 2-4 cells.
Fibres non-septate, thin-walled (2-2.5 µm), diam. 23-25 µm. Intercellular cavities obvious. Pits minute, bordered, in radial walls. Length of fibres 1272 (990-1450) µm. F/V ratio 2.2.

Material studied (Uw-numbers refer to the Utrecht Wood collection): **H. guianensis** Aubl.: Suriname: Stahel 380 = Uw 380; Lanjouw & Lindeman 1547 = Uw 1509.

SPARATTANTHELIUM Mart. Fig. 59 B-D

Growth rings distinct, boundaries indicated by a few rows of flattened fibres and by an almost continuous band of aliform/confluent parenchyma.
Vessels for 60-70% solitary, the remainder in radial pairs or in multiples of 3-4, occasionally in small clusters; outline round to elliptic, medium-sized to large, mostly 100-220 µm, but up to 300 µm in *S. wonotoboense*; number on average 5 (3-7) per mm^2. Perforations simple. Vessel member length 370 (200-650) µm. Intervessel pits 10-14 µm, apertures included. Vessel-ray pits large, oval, more or less in scalariform arrangement, the borders strongly reduced.
Rays 1-3-seriate, mainly biseriate; 3 (2-7) per mm; almost homocellular, marginal cells radially slightly shorter, but of the same height as the procumbent cells; width 20-50 µm, height up to 235 µm (30 cells).

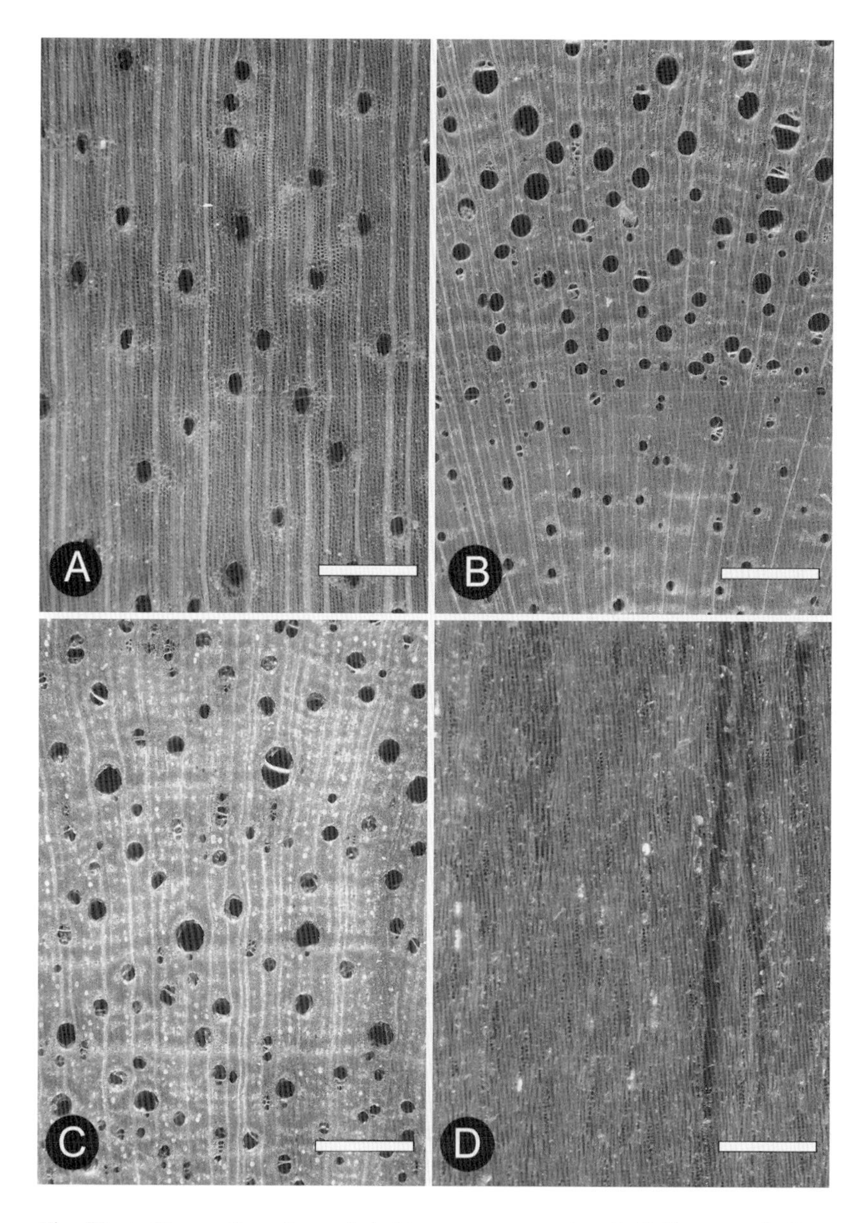

Fig. 59. A, *Hernandia guianensis* Aubl.: transverse section; B-C, *Sparattanthelium wonotobense* Kosterm.: transverse sections; D, *Sparattanthelium guianense* Sandwith: tangential section (scale bars A-C = 1 mm; D = 0.5 mm).

Parenchyma abundant, chiefly paratracheal, vasicentric to aliform and aliform-confluent but also some apotracheal diffuse and diffuse-in-aggregates present; strands of 2-4 cells. Large cystoliths in inflated cells numerous in all species except *S. wonotoboense*.
Fibres non-septate, thin-walled with a large lumen, diam. 20-23 µm. Pits minute, simple, mainly in radial walls. Length of fibres 860 (750-970) µm. F/V ratio 2.2.

Material studied (Uw-numbers refer to the Utrecht Wood collection):
S. aruakorum Tutin: Suriname: van Donselaar 3754 = Uw 12148.
S. guianense Sandwith: Guyana: Fanshawe 3938 = Uw 24087.
S. wonotoboense Kosterm.: Guyana: A.C.Smith 3390 = Uw 24191; Jansen-Jacobs *et al.* 3180 = Uw 34575; Pipoly *et al.* 7437 = Uw 32739.

PIPERACEAE

by

JIFKE KOEK-NOORMAN[5]

WOOD ANATOMY

The family consists of mainly herbaceous or (sub)shrubby species. Only in *Piper*, larger trees or climbing species occur. However, the stem diameter rarely exceeds a few cm (for instance: the available sample of the climbing species *P. poiteanum* was less than 1 cm in diam.). The yellowish to brown wood of Piperaceae is characterized by a combination of small vessels and broad to very broad and high rays consisting of upright and square cells, forming up to 50 % of the surface in transverse section and often continuing from pith to bark.
As far as known the wood is of no commercial value.

GENERIC DESCRIPTION

PIPER L. – Figs. 60-61

Vessels round to oval, solitary and in small multiples, diffuse, regularly distributed, diameter 30-45 µm in *P. bartlingianum*, to 100-150 µm in *P. aduncum*, often small and large vessels intermingled; number per mm^2 often difficult to count because of the relatively narrow zones between the broad rays, from 5-11 in *P. aduncum* to 36-50 in *P. poiteanum*. Perforations simple. Intervascular pits oval, 5 x 6 µm, with anastomosing apertures (*P. aduncum, P. poiteanum,* and *P. reticulatum*) or scalariform (*P. arboreum, P. bartlingianum, P. vs. crassinervium, P. hostmannianum, P. obliquum,* and *P. peltata*). Vessel-parenchyma pits elongate to scalariform. Vessel member length 150-200 µm (*P. aduncum*) to more than 500 µm (*P.* vs. *crassinervium* and *P. obliquum*). Some wood samples show a slight tendency to storied structure due to the axial arrangement of the vessel members and parenchyma strands.
Rays very high and wide, up to ca. 15-25 cells, 200-450 µm broad (but ca. 9 cells, 100 µm in *P. bartlingianum*), consisting of slightly irregular, square and upright cells; in *P. aduncum* and in one sample of *P. obliquum*, islands of procumbent cells are included. Empty ideoblasts in *P. aduncum.* In *P. hostmannianum* the outer 2.5 mm is unlignified. Due

[5] Nationaal Herbarium Nederland, Utrecht University branch, Heidelberglaan 2, 3584 CS Utrecht, The Netherlands.

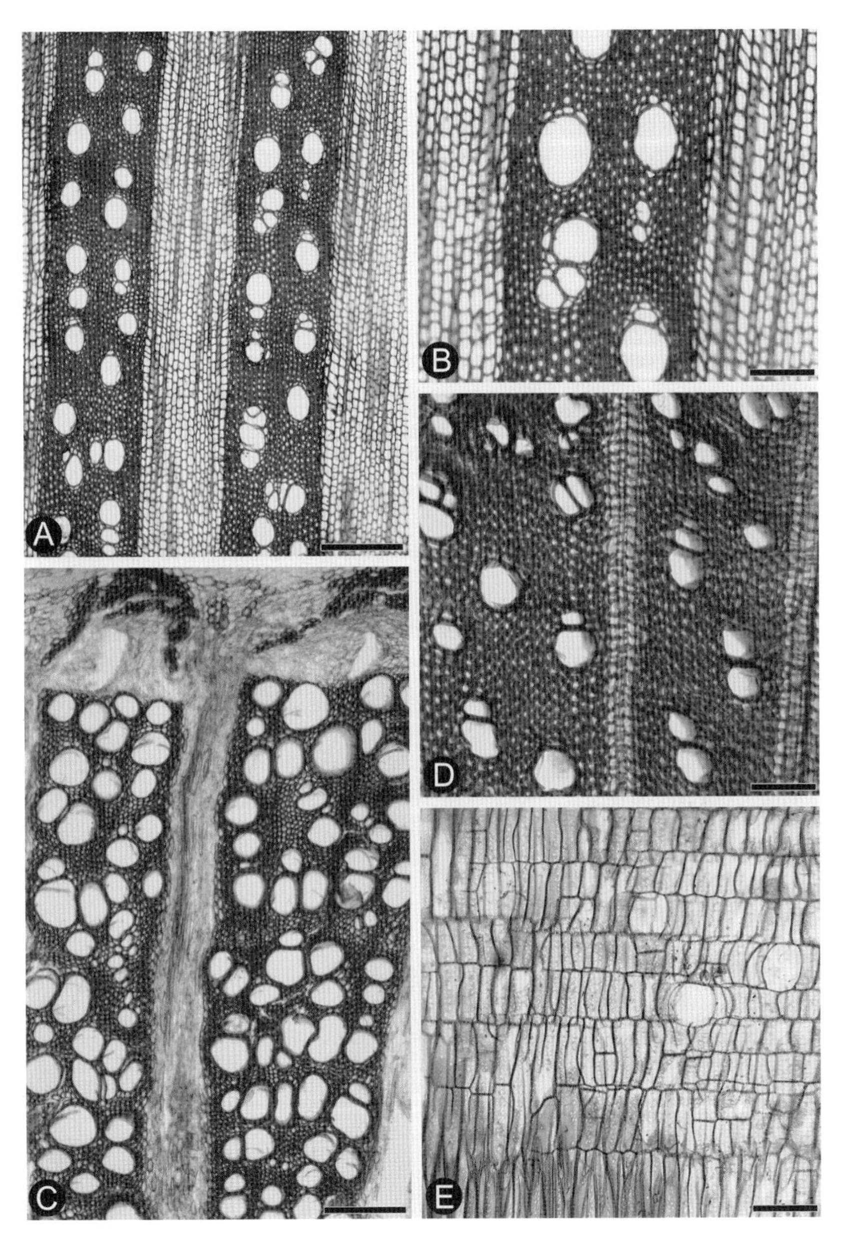

Fig. 60. Transverse sections of *Piper*: A, *Piper arboreum* Aubl.: vessel distribution; B, id.: paratracheal parenchyma; C, *Piper hostmannianum* (Miq.) C. DC.: rays towards the bark unlignified; D, *Piper peltatum* L.: vessel distribution; E, *Piper aduncum* L.: radial section, showing ideoblasts in rays (scale bars A, C = 350 μm; B, D = 100 μm).

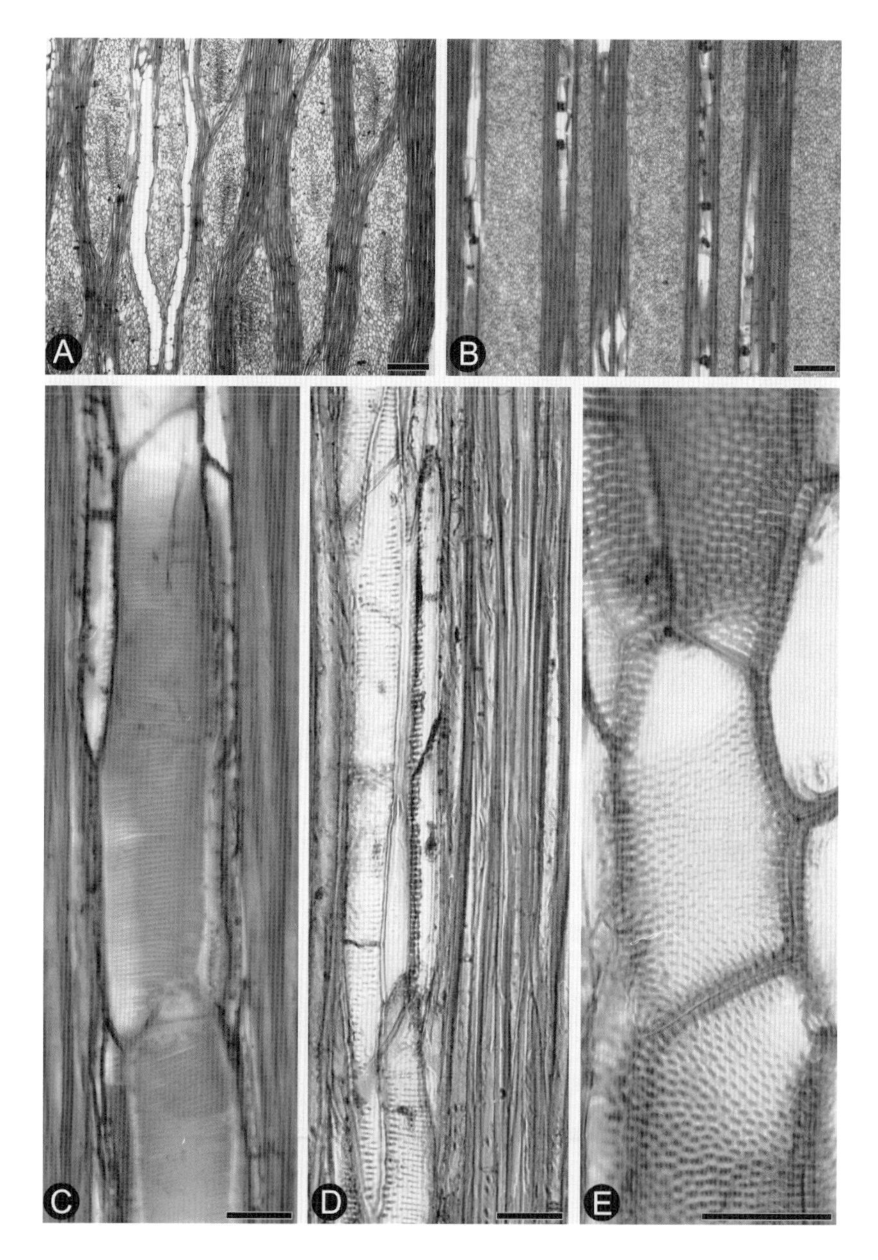

Fig. 61. Tangential sections of *Piper*. A, *Piper aduncum* L.: islands of procumbent cells in rays; B, *Piper reticulatum* L.: very high and broad rays; coloured inclusions in vessels; C, *Piper arboreum* Aubl.: intervessel pitting; D, id.: vessel-parenchyma pitting; E, *Piper aduncum* L.: intervessel pitting (scale bars A-B = 350 μm; C-E = 35 μm).

to the structure of the wood, vessel-ray pits could hardly be found. The scanty pits that could be observed, were similar to the vessel-parenchyma pits.
Parenchyma paratracheal-vasicentric in narrow complete or incomplete rings; fusiform cells or 2(-4)-celled strands.
Fibres moderately thick-walled, walls 3-4 µm, lumina up to 15 µm wide, sometimes septate. Pits simple to minutely bordered, slitlike, numerous on radial walls, scanty on tangential walls.

Remarks: My findings agree with the data found in the literature, which however are scanty due to the fact that the wood is of no commercial value. In general, the wood is rather homogeneous. The main difference was found in the intervessel pitting and the vessel-parenchyma pitting which vary from alternate-confluent to scalariform. No correlation was found with other, quantitative characters like vessel diameter and number, vessel member length, or ray type.
P. bartlingianum shows the narrowest vessels and the narrowest rays.
The ideoblasts found in the rays of *P. aduncum* remind of Williams' remark (1936) that the wood is sometimes fragrant.
The transverse section of *P. hostmannianum* is highly similar to *P. poiteanum*, the only climbing species studied, viz. in vessel arrangement, and in the rays, which are unlignified towards the phloem. However, differences are found in the vessel pits, those of *P. poiteanum* being alternate to confluent, and in the ray cells, those of *P. poiteanum* being square (to very slightly upright).
P. peltatum, considered long-time in the *Pothomorphe* genus fits nicely in *Piper*.

Material studied (Uw-numbers refer to the Utrecht Wood collection):
Piper aduncum L.: Suriname: Lanjouw & Lindeman 1801 = Uw 1553.
P. aequale Vahl: Guyana: Jansen-Jacobs *et al.* 324 = Uw 30585.
P. arboreum Aubl.: Suriname: Daniëls & Jonker 1171 = Uw 8623.
P. augustum Rudge: Suriname: Irwin *et al.* 54677 = Uw 17397; Guyana: Maas *et al.* 5601 = Uw 27312.
P. avellanum (Miq.) C. DC.: Suriname: Heyde 276 = Uw 23321.
P. bartlingianum (Miq.) C.DC.: Guyana: Jansen-Jacobs *et al.* 2333 = Uw 33970.
P. vs. **crassinervium** Kunth: Suriname: Irwin *et al.* 54816 = Uw 17410.
P. hispidum Sw.: French Guiana: de Granville *et al.* 7448 = Uw 31435; Guyana: Jansen-Jacobs *et al.* 399 = Uw 30616.
P. hostmannianum (Miq.) C. DC.: Guyana: Maas *et al.* 5412 = Uw 27256.
P. insipiens Trel. & Yunck.: French Guiana: de Granville *et al.* 6561 = Uw 30019; Suriname: Irwin *et al.* 54654 = Uw 17322.
P. marginatum Jacq.: Suriname: Lindeman & Heyde 336 = Uw 22884.

P. obliquum Ruiz & Pav.: Suriname: Daniëls & Jonker 979 = Uw 8571; Guyana: Jansen-Jacobs *et al.* 6035 = Uw 36651; Maas *et al.* 5856 = Uw 27363.
P. peltatum L.: Guyana: Jansen-Jacobs *et al.* 434 = Uw 30637.
P. poiteanum Steud.: Suriname: Heyde 275 = Uw 23320.
P. reticulatum L.: Suriname: Lindeman & de Roon 801 = Uw 27449; Guyana: Jansen-Jacobs *et al.*1059 = Uw 32207.

LITERATURE ON WOOD AND TIMBER

Carlquist, S. 1962a. A theory of paedomorphosis in dicotyledonous woods. Phytomorphology 12: 30-45.

Dadswell, H.E. & S.J. Record. 1936. Identification of woods with conspicuous rays. Trop Woods 48: 1-30.

Dechamps, R. 1985. Etude anatomique de bois d' Amérique du Sud. III. Linaceae-Quiinaceae. Ann. Mus. roy. Afr. Cent. Tervuren, ser. IN-8, Sci. econ., No 14: 471 pp.

Détienne, P. & P. Jacquet.1983. Atlas d'identification des bois de l'Amazonie et des régions voisines. C.T.F.T., Nogent sur Marne.

Ilic, J. 1991. CSIRO atlas of hardwoods. Springer-Verlag, Berlin, etc. 525 pp.

Kribs, D.A. 1968. Commercial Foreign Woods on the American Market. Dover Publ., New York.

Lindeman, J.C., A.M.W. Mennega & W.H.A. Hekking. 1963. Bomenboek voor Suriname. Dienst 's Lands Bosbeheer, Paramaribo.

Metcalfe, C.R. 1987. Anatomy of the Dicotyledons, vol. 3, sec. ed. Oxford Univ. Press.

Metcalfe, C.R. & L. Chalk. 1950. Anatomy of the Dicotyledons, Volume 2. Clarendon Press, Oxford.

Record, S.J. & R.W. Hess. 1943. Timbers of the New World. Yale school of Forestry: New Haven. 640 pp.

Shutts, C.F. 1960. Wood anatomy of Hernandiaceae and Gyrocarpaceae. Trop. Woods 113.

Welle, B.J.H. ter, 1980. Cystoliths in the secondary xylem of Sparattanthelium (Hernandiaceae). IAWA Bull. n.s. Vol. I.

Williams, L. 1936. Woods of Northeastern Peru. Field Mus. nat. Hist., Bot. ser. 15: 1-187. Chicago.

TAXONOMIC CHANGES

The following taxon are newly placed in synonymy:

Hernandiaceae

Sparattanthelium botocudorum Mart. var. *uncigerum* Meisn. to Sparattanthelium wonotoboense Kosterm.
Sparattanthelium uncigerum (Meisn.) Kubitzki to S. wonotoboense Kosterm.

Piperaceae

Artanthe schomburgkii Klotzsch to Piper guianense (Klotzsch) C. DC.
Peperomia muscosa Link to Peperomia quadrangularis (J.V. Thomps.) A. Dietr.
Piper amapense Yunck. to Piper inaequale C. DC.
Piper fockei Trel. & Yunck. to Piper paramaribense C. DC.
Piper gabrielianum (Miq.) C. DC. to Piper avellanum (Miq.) C. DC.
Piper gleasonii Yunck. to Piper trichoneuron (Miq.) C. DC.
Piper gleasonii Yunck. var. *wonotoboense* Yunck. to Piper hostmannianum (Miq.) C. DC.
Piper insigne (Kunth) Steud. to Piper obliquum Ruiz & Pav.
Piper kappleri C. DC. to Piper cyrtopodum (Miq.) C. DC.
Piper kegelianum (Miq.) C. DC. to Piper trichoneuron (Miq.) C. DC.
Piper lenormandianum C. DC. to Piper demeraranum (Miq.) C. DC.
Piper liesneri Steyerm. to Piper brasiliense C. DC.
Piper maguirei Yunck. to Piper paramaribense C. DC.
Piper nematanthera C. DC. to Piper hymenophyllum (Miq.) Wight
Piper oblongifolium (Klotzsch) C. DC., var. *glabrum* C. DC. to Piper guianense (Klotzsch) C. DC.
Piper romboutsii Yunck. to Piper divaricatum G. Mey.
Piper subciliatum C. DC. to Piper cyrtopodum (Miq.) C. DC.

NUMERICAL LIST OF ACCEPTED TAXA

Hernandiaceae

1. Hernandia L.
    - 1-1. H. guianensis Aubl.

2. Sparattanthelium Mart.
    - 2-1. S. aruakorum Tutin
    - 2-2. S. guianense Sandwith
    - 2-3. S. wonotoboense Kosterm.

Chloranthaceae

1. Hedyosmum Swartz
    - 1-1. H. tepuiense Todzia

Piperaceae

1. Peperomia Ruiz & Pav.
    - 1-1. P. alata Ruiz & Pav.
    - 1-2. P. angustata Kunth
    - 1-3. P. blanda (Jacq.) Kunth
    - 1-4. P. delascioi Steyerm.
    - 1-5. P. emarginella (Sw. ex Wikstr.) C. DC.
    - 1-6. P. galioides Kunth
    - 1-7. P. glabella (Sw.) A. Dietr.
    - 1-8. P. gracieana Görts
    - 1-9. P. haematolepis Trel.
    - 1-10. P. hernandiifolia (Vahl) A. Dietr.
    - 1-11. P. lancifolia Hook. subsp. lancifolia
    - 1-12. P. macrostachya (Vahl) A. Dietr.
    - 1-13. P. magnoliifolia (Jacq.) A. Dietr.
    - 1-14. P. maguirei Yunck.
    - 1-15. P. obtusifolia (L.) A. Dietr.
    - 1-16. P. ouabianae C. DC.
    - 1-17. P. pellucida (L.) Kunth
    - 1-18. P. pernambucensis Miq.
    - 1-19. P. popayanensis Trel. & Yunck.
    - 1-20. P. purpurinervis C. DC.
    - 1-21. P. quadrangularis (J.V. Thomps.) A. Dietr.
    - 1-22. P. quadrifolia (L.) Kunth

1-23. P. reptans C. DC.
1-24. P. rhombea Ruiz & Pav.
1-25. P. rotundifolia (L.) Kunth
1-26. P. serpens (Sw.) Loudon
1-27. P. tenella (Sw.) A. Dietr.
1-28. P. tetraphylla (G. Forstr.) Hook. & Arn.
1-29. P. transparens Miq.
1-30. P. urocarpa Fisch. & Mey.

2. Piper L.
2-1. P. adenandrum (Miq.) C. DC.
2-2. P. aduncum L.
2-3. P. aequale Vahl
2-4. P. alatabaccum Trel. & Yunck.
2-5. P. amalago L.
2-6. P. angustifolium Lam.
2-7. P. anonifolium (Kunth) Steud.
2-7a. var. anonifolium
2-7b. var. parkerianum (Miq.) Steyerm.
2-8. P. arboreum Aubl.
2-9. P. augustum Rudge
2-10. P. aulacospermum Callejas
2-11. P. avellanum (Miq.) C. DC.
2-12. P. bartlingianum (Miq.) C. DC.
2-13. P. bolivaranum Yunck.
2-14. P. brasiliense C. DC.
2-15. P. brownsbergense Yunck.
2-16. P. cernuum Vell.
2-17. P. ciliomarginatum Görts & Christenhusz
2-18. P. consanguineum (Kunth) Steud.
2-19. P. coruscans Kunth
2-20. P. crassinervium Kunth
2-21. P. cyrtopodum (Miq.) C. DC.
2-22. P. demeraranum (Miq.) C. DC.
2-23. P. dilatatum Rich.
2-24. P. divaricatum G. Mey.
2-25. P. duckei C. DC.
2-26. P. dumosum Rudge
2-27. P. eucalyptifolium Rudge
2-28. P. fanshawei Yunck.
2-29. P. flexuosum Rudge
2-30. P. fuligineum (Kunth) Steud.
2-31. P. glabrescens (Miq.) C. DC.
2-32. P. guianense (Klotzsch) C. DC.

2-33. P. hirtilimbum Trel. & Yunck.
2-34. P. hispidum Sw.
2-35. P. hostmannianum (Miq.) C. DC.
2-36. P. humistratum Görts & K.U. Kramer
2-37. P. hymenophyllum (Miq.) Wight
2-38. P. inaequale C. DC.
2-39. P. insipiens Trel. & Yunck.
2-40. P. marginatum Jacq.
2-41. P. nigrispicum C. DC.
2-42. P. nigrum L.
2-43. P. obliquum Ruiz & Pav.
2-44. P. paramaribense C. DC.
2-45. P. peltatum L.
2-46. P. perstipulare Steyerm.
2-47. P. phytolaccifolium Opiz
2-48. P. piscatorum Trel. & Yunck.
2-49. P. poiteanum Steud.
2-50. P. pulleanum Yunck.
2-51 P. remotinervium Görts
2-52. P. reticulatum L.
2-53. P. rudgeanum (Miq.) C. DC.
2-54. P. rupununianum Trel. & Yunck.
2-55. P. salicifolium Vahl
2-56. P. trichoneuron (Miq.) C. DC.
2-57. P. tuberculatum Jacq.
2-58. P. wachenheimii Trel.

COLLECTIONS STUDIED
(Numbers in bold represent types)

Hernandiaceae

GUYANA

Andel, T. van *et al.*, 2646 (1-1)
Anderson, C.W., 12 (1-1)
Clarke, D. *et al.*, 6178 (2-3)
Cruz, J.S. de la, 1395, 3093 (1-1)
Forest Dept. British Guiana (FD), 5586=Fanshawe 2787 (2-1); 3355=Fanshawe 619 (1-1); 3938=Fanshawe 1202 (2-2)
Gillespie, L.J. & H. Persaud, 1202 (1-1)
Hitchcock, A.S., 17622 (1-1)
Hoffman, B. *et al.*, 2764 (2-2)
Jansen-Jacobs, M.J. *et al.*, 3180, 3557, 4488 (2-3)
Jenman, G.S., 4889 (2-2)
Maas, P.J.M. *et al.*, 5562 (1-1)
Sandwith, N.Y., **470** (2-2)
Smith, A.C., **3390** (2-3)
Snow, D.W.,14 (2-3)
Pipoly, J.J. *et al.*, 7437 (2-3)
Polak, M. *et al.*, 72 (1-1)
Tutin, T.G., **252** (2-1)

SURINAME

BBS 112 (1-1)
BW 802 (1-1); **3120**, 6559 (2-3)
Collector indigenus surinamensis, 164 (2-3)
Lanjouw, J. & J.C. Lindeman, 1547 (1-1)
Lindeman, J.C. & A.R.A. Görts-van Rijn *et al.*, 57, 363 (2-3)
Maguire, B., 24127 (2-3)
Mennega, A.M.W., 463 (2-3)
Stahel, G. & J.W. Gonggrijp, in BW 802 (1-1)

FRENCH GUIANA

Aublet, J.B.C.F., **s.n**. (1-1)
Béna, P., in BAFOG 47N (1-1)
Cremers, G., 7020 (1-1); 9581 (2-3)
Feuillet, C., 239 (2-3)
Granville J.J. de, 2811, 6900 (1-1)
Grenand, P., 2101 (2-3)
Hallé, F., 4019 (2-3)
Jacquemin, H., 2790 (1-1)
Kubitzki, K., 71-116 (1-1)
Leeuwenberg, A.J.M., 11781 (1-1)
Mélinon, M., s.n. (2-3)
Mori, S.A. *et al.*, 18986, 23044, 23571 (2-3)
Oldeman, R.A.A, B-1029 (1-1); B-3890 (2-3)
Prévost, M.F., 1512 (2-3); 3438, 3500, 4085 (1-1); 4096 (2-3)
Riera, B., 792 (1-1)
Sagot, P., **1218** (2-3)

Chloranthaceae

GUYANA

Boom, B.M. *et al.*, 9063, 9200 (1-1)
Hahn, W.J. *et al.*, 5481a (1-1)

Piperaceae

GUYANA

Acevedo, P., 3305 (1-21); 3353 (2-8); 3370 (2-34); 3382 (2-11); 3415 (1-12)
Altson, R.A., 96 (2-34); 482 (2-31)
Andel, T. van, 737 (2-7a); 1103 (2-35); 1250 (2-11); 1264 (2-3); 1333 (1-25); 1353 (2-8); 1469 (2-41); 1524 (2-1); 1586 (2-35); 1667 (2-43); 1695 (2-52); 1795 (2-31); 1800 (2-11); 1949 (1-15); 1986, 2108 (2-8); 2115 (2-11); 2181 (2-8); 2189 (2-11); 2201 (2-1); 2265 (1-15); 2430 (2-36); 2473 (2-1); 2567 (2-35); 3013 (2-41); 3014 (2-35); 3023 (1-7)
Anon. Botanic Gardens, Ao 1909 (1-25)
Appun, C.F., 398 (2-35); 649 (1-16); 747 (1-17); Ao 1872 (1-12); Ao 1872 (1-25)
Archer, W.A., 25 (2-34); 2299 (2-23); 2300 (2-45); 2362 (2-8); 2534 (2-31); 2535 (2-8); **2505** (2-19); 2518 (2-34)
Bartlett, A.W., 243 (2-56); 244 (2-11); 8211 (1-26); **8233** (1-25); 8252 (1-7); 8408 (2-54); 8581, 8741, 8742 (1-12); 8743 (2-9); 8744 (1-12); 8748, 8749 (1-12); 8759 (2-8); 8761 (1-7); 8812 (2-35); s.n. (2-2); s.n. (2-22); Ao 1905 (1-26); Ao 1907 (2-2); Ao 1908 (1-26)
Beckett, J.E., 8340 (1-5); 8630, 8631 (2-58); s.n. (1-25); s.n. (2-7a)
Boom, B., 7322, 8426 (2-35)
Campbell, W.H., (Coll. Drake 6/71) s.n. (2-24); (Coll. Drake 6/71) s.n. (1-25);
Chanderbali, A. *et al.*, 45 (2-35); 104 (2-34); 398, 535 (2-12); 543 (1-21); 570 (1-12); 603 (1-13)
Clarke, D., 5 (2-12); 178 (2-56); 232 (2-35); 316 (2-12); 404 (2-35); 434 (2-56); 541 (2-11); 559, 595 (1-12); 598 (2-7a); 761 (2-35); 797 (1-12); 834 (2-35); 949 (1-12); 964 (2-22); 968 (2-2); 1076 (2-8); 1103 (2-45); 1147 (1-30); 1175 (1-16); 1191 (1-12); 1198 (2-7a); 1218 (2-34); 1219 (2-8); 1289 (2-12); 1312 (2-35); 1313 (2-22); 1314 (2-12); 1346, 1494 (1-12); 1618 (2-35); 1745 (2-34); 1747 (1-21); 2035, 2259 (2-12); 2268 (1-12); 2329, 2405 (2-12); 2489 (2-39); 2497, 2513 (2-8); 2799 (1-25); 2805 (2-14); 2808 (2-18); 2809 (1-3); 2810 (1-12); 2886 (2-4); 2902 (2-12); 2934 (2-9); 2935 (2-34); 2946 (2-4); 2947 (2-1); 3059 (1-7); 3106 (2-20); 3135 (2-33); 3143 (2-18); 3169 (2-4); 3179 (2-33); 3232 (2-4); 3299 (1-12); 3351 (2-4); 3402 (1-12); 3425 (2-12); 3438 (1-12); 3450 (1-17); 3609 (2-12); 3630 (2-8); 3639 (1-25); 3663 (1-12); 3724 (2-15); 3756 (1-12); 3768 (1-7); 3775 (1-12); 3884 (2-12); 3890 (2-7a); 3951 (2-37); 4003 (1-12); 4066 (2-56); 4128 (1-16); 4144 (2-39); 4176 (1-13); 4179 (2-39); 4255 (1-26); 4283 (2-34); 4294 (2-22); 4306 (1-16);

SURINAME

Splitgerber, F.L., 38 (1-17); 41 (2-34); 63 (1-25); 142 (2-45); 181 (2-24); 225 (2-23); 226 (2-8); 227 (2-24); 238 (1-12); 301 (1-21); 339 (1-13); 492 (2-18); 1138 (1-26); 1139, s.n. (2-2); s.n. (2-24); s.n. (2-34)

Stahel, G., 67 (2-12); 392 (1-13)

Stoffers, A.L., A.R.A. Görts-van Rijn *et al.*, 185 (2-34); 227 (2-52)

Suringar, W.F.R., s.n. (2-1)

Tjon Lim Sang, R. & I. v.d. Wiel, 62 (2-36); 91 (2-8); 92, 107a+b, 113 (2-34); 133 (2-3); 174 (2-8); 186 (2-34); 194 (2-35); 195, 199 (2-34); 200 (2-8); 221 (2-3)

Tresling, O., 22 (2-35); 164 (1-26); 167 (1-17); 206 (2-45); 238 (1-13); 420 (1-12); 474 (1-12); 478 (1-25); 480 (1-7)

Troon, F. van, CI 325 (2-2)

Tulleken, J.E., 15, 103 (1-17); 176 (2-40); 212 (2-56); 258 (1-17); 285 (2-45); 506 (2-7a); 573 (2-24); s.n. (1-17)

Unknown collector, 4 (2-34); s.n. (1-12); s.n. (1-25); s.n. (1-26); s.n. (2-2); s.n. (2-8); s.n. (2-34); s.n. (2-40); s.n. (2-45)

UVS, 16586 (1-17); 16710 (1-14); 16965 (1-17); 16968 (2-40); 17197 (1-17); 17199 (2-40); 17751 (2-12); 17807 (1-15); 17815 (2-58); 17835 (2-6); 17851 (1-25); 17852 (2-35); s.n. (2-42)

Versteeg, G.M., 102 (2-24); 185 (2-18); 195 (1-12); 336 (2-9); 341 (2-3); 363 (1-21); 666 (1-17); 816 (2-24)

Veth, B., 115 (2-8)

Webster, G.L., 24031 (2-35); 24089 (2-4); 24121 (2-8)

Weigelt, C., 1821, s.n. (2-2); s.n. (2-8); s.n. (2-40)

Went, F.A.F.C., 13 (1-17); 71 (2-22); 89 (2-23); 162 (1-7); 201, 312 (2-34); 361 (2-24); 513 (2-8); 541 (2-24)

Wessels Boer, J.G., 317 (2-12); 318 (2-22); 447 (1-25); 816 (2-56); 843 (2-30); 853 (2-18); 951, 957 (1-26); 965 (2-50); 987 (1-12); 1023 (2-8); 1058 (2-3); 1095 (2-9); 1098 (2-18); 1193 (2-12); 1238 (1-16); 1269 (2-34); 1294 (2-36); 1399 (2-12); 1427 (1-13); **1469** (2-36); 1533 (2-18); 1599 (2-3)

Wilhelmina Mts. Exp., 157 (2-45)

FRENCH GUIANA

Alexandre, D.Y., 449 (2-6)

Allorge, L., 68 (2-44); 69 (2-18); 125 (2-1); 131 (2-7)

Aublet, J.B.C.F. d', **s.n.** (2-8); s.n. (2-34); **s.n.** (2-36)

Aubréville, A., 38 (2-2); 148 (2-35)

Aumeeruddy, Y., 53 (2-35); 54 (2-58); 59 (2-3); 64 (2-5); 96 (2-12); 113 (2-35)

BAFOG, 7036 (2-2); 7970 (2-34)

Bahri, S., 19 (2-35); 25 (2-36)

Barabé, L., 116 (2-35)

Barthelemy, D., 128 (2-18); 130 (2-36); 132 (2-8); 140 (2-34); 144 (2-1); 146 (1-18); 148 (2-3); 149 (2-8); 151 (2-23)

INDEX TO SYNONYMS AND SPECIES MENTIONED IN NOTES

Hernandiaceae

Piperaceae

INDEX TO VERNACULAR NAMES

Hernandiaceae

Piperaceae

Alphabetic list of families of series A occurring in the Guianas

Defined as in Cronquist, 1981, and numbered in his sequence, with alternative names. Those published, with chronological fascicle number and year.

Abolbodaceae		
(see Xyridaceae	182)	15. 1994
Acanthaceae	156	23. 2005
(incl. Thunbergiaceae)		
(excl. Mendonciaceae	159)	
Achatocarpaceae	028	22. 2003
Agavaceae	202	
Aizoaceae	030	22. 2003
(excl. Molluginaceae	036)	22. 2003
Alismataceae	168	
Amaranthaceae	033	22. 2003
Amaryllidaceae		
(see Liliaceae	199)	
Anacardiaceae	129	19. 1997
Anisophylleaceae	082	
Annonaceae	002	
Apiaceae	137	
Apocynaceae	140	
Aquifoliaceae	111	
Araceae	178	
Araliaceae	136	
Arecaceae	175	
Aristolochiaceae	010	20. 1998
Asclepiadaceae	141	
Asteraceae	166	
Avicenniaceae		
(see Verbenaceae	148)	4. 1988
Balanophoraceae	107	14. 1993
Basellaceae	035	22. 2003
Bataceae	070	
Begoniaceae	065	
Berberidaceae	016	
Bignoniaceae	158	
Bixaceae	059	
(incl. Cochlospermaceae)		
Bombacaceae	051	
Bonnetiaceae		
(see Theaceae	043)	
Boraginaceae	147	
Brassicaceae	068	
Bromeliaceae	189	p.p. 3. 1987
Burmanniaceae	206	6. 1989
Burseraceae	128	
Butomaceae		
(see Limnocharitaceae	167)	
Byttneriaceae		
(see Sterculiaceae	050)	
Cabombaceae	013	
Cactaceae	031	18. 1997
Caesalpiniaceae	088	p.p. 7. 1989
Callitrichaceae	150	
Campanulaceae	162	
(incl. Lobeliaceae)		
Cannaceae	195	1. 1985
Canellaceae	004	
Capparaceae	067	
Caprifoliaceae	164	
Caricaceae	063	
Caryocaraceae	042	
Caryophyllaceae	037	22. 2003
Casuarinaceae	026	11. 1992
Cecropiaceae	022	11. 1992
Celastraceae	109	
Ceratophyllaceae	014	
Chenopodiaceae	032	22. 2003
Chloranthaceae	008	24. 2006
Chrysobalanaceae	085	2. 1986
Clethraceae	072	
Clusiaceae	047	
(incl. Hypericaceae)		
Cochlospermaceae		
(see Bixaceae	059)	
Combretaceae	100	
Commelinaceae	180	
Compositae		
(= Asteraceae	166)	
Connaraceae	081	
Convolvulaceae	143	
(excl. Cuscutaceae	144)	
Costaceae	194	1. 1985
Crassulaceae	083	
Cruciferae		
(= Brassicaceae	068)	
Cucurbitaceae	064	
Cunoniaceae	081a	
Cuscutaceae	144	
Cycadaceae	208	9. 1991
Cyclanthaceae	176	
Cyperaceae	186	
Cyrillaceae	071	
Dichapetalaceae	113	
Dilleniaceae	040	
Dioscoreaceae	205	
Dipterocarpaceae	041a	17. 1995
Droseraceae	055	22. 2003
Ebenaceae	075	

Piperaceae	009	24. 2006
Plantaginaceae	151	
Plumbaginaceae	039	
Poaceae	187	8. 1990
Podocarpaceae	211	9. 1991
Podostemaceae	091	
Polygalaceae	125	
Polygonaceae	038	
Pontederiaceae	197	15. 1994
Portulacaceae	034	22. 2003
Potamogetonaceae	171	
Proteaceae	090	
Punicaceae	097	
Quiinaceae	045	
Rafflesiaceae	108	
Ranunculaceae	015	
Rapateaceae	181	
Rhabdodendraceae	086	
Rhamnaceae	116	
Rhizophoraceae	101	
Rosaceae	084	
Rubiaceae	163	
(incl. Henriquesiaceae)		
Ruppiaceae	172	
Rutaceae	132	
Sabiaceae	018	
Santalaceae	104	
Sapindaceae	127	
Sapotaceae	074	
Sarraceniaceae	054	22. 2003
Scrophulariaceae	153	
Simaroubaceae	130	
Smilacaceae	204	
Solanaceae	142	
Sphenocleaceae	161	
Sterculiaceae	050	
(incl. Byttneriaceae)		
Strelitziaceae	190	1. 1985
Styracaceae	076	
Suraniaceae	086a	
Symplocaceae	078	
Taccaceae	203	
Tepuianthaceae	114	
Theaceae	043	
(incl. Bonnetiaceae)		
Theophrastaceae	079	
Thunbergiaceae		
(see Acanthaceae	156)	
Thurniaceae	185	
Thymeleaceae	095	
Tiliaceae	049	17. 1995
Trigoniaceae	124	21. 1998
Triuridaceae	174	5. 1989
Tropaeolaceae	135	
Turneraceae	061	
Typhaceae	188	
Ulmaceae	020	11. 1992
Umbelliferae		
Urticaceae	023	11. 1992
Valerianaceae	165	
Velloziaceae	201	
Verbenaceae	148	4. 1988
(incl. Avicenniaceae)		
Violaceae	060	
Viscaceae	106	
Vitaceae	117	
Vochysiaceae	123	21.1998
Winteraceae	001	
Xyridaceae	182	15. 1994
(incl. Albolbodaceae)		
Zamiaceae	208a	9. 1991
Zingiberaceae	193	1. 1985
(excl. Costaceae	194)	
Zygophyllaceae	133	